Chemistry and Chemical Engineering in the Chlor-alkali Industry: Key Components and Implications

CHEMISTRY AND CHEMICAL ENGINEERING IN THE CHLOR-ALKALI INDUSTRY: KEY COMPONENTS AND IMPLICATIONS

Edited by:

Dr. Saeed Farrokhpay

ARCLER PRESS

www.arclerpress.com

Chemistry and Chemical Engineering in the Chlor-alkali Industry: Key Components and Implications

Dr. Saeed Farrokhpay

Arcler Press
224 Shoreacres Road
Burlington, ON L7L 2H2
Canada
www.arclerpress.com
Email: orders@arclereducation.com

ISBN: 978-1-77469-821-1 (Hardcover)

Arcler Press publishes wide variety of books and eBooks. For more information about Arcler Press and its products, visit our website at www.arclerpress.com

ABOUT THE EDITOR

Dr Saeed Farrokhpay is a Chemical Engineer with several years of experience in mineral & material processing. He obtained his PhD from University of South Australia in 2005. He is currently a Technical Consultant in Australia. He has worked for more than 20 years at mineral and chemical industries, universities and research centers around the world. Dr Farrokhpay has published more than 90 papers in high ranked journals and conference proceedings. He has also edited several technical and scientific books, and served as an editorial board member of several international scientific journals.

TABLE OF CONTENTS

LIST OF FIGURES

LIST OF TABLES

.

LIST OF ABBREVIATIONS

AOXs	Adsorbable Organic Halides
BAT	Best Available Technology
BREFs	BAT Reference Documents
BRR	Benefit-to-Risk Ratio
CAER	Community Awareness and Emergency Response
CCl_4	Carbon Tetrachloride
CFC-11	Chlorofluorocarbon-11
CFCs	Chlorofluorocarbons
$CHCl_3$	Chloroform
CO_2	Carbon Dioxide
DSAs	Dimensionally Stable Anodes
EBCHM	Electron Beam Cold Hearth Melting
ECU	Electrical Unit
EDC	Ethylene Dichloride
ELVs	Emission Limit Values
EMF	Electromotive Force
EPA	Environmental Protection Agency
EPI	Epichlorohydrin
GWP	Global Warming Potential
HCFC-22	Chlorodifluoromethane
HCFCs	Hydrochlorofluorocarbons
HER	Hydrogen Evolution Reaction
ICG	Inter-Sessional Correspondence Group
IEF	Information Exchange Forum
KOH	Potassium Hydroxide

LLE	Loss of Life Expectancy
MDI	Methylene Diphenyl Diisocyanate
$Mg(OH)_2$	Magnesium Hydroxide
MMO	Mixed Metal Oxide
NPVs	Net Present Values
ODP	Ozone Depletion Potential
OH^-	Hydroxides
PAM	Plasma Arc Melting
PCE	Perchloroethylene
PO	Propylene Oxide
POPs	Persistent Organic Pollutants
PTFE	Polytetrafluoroethylene
PVC	Polyvinyl Chloride
RHE	Reversible Hydrogen Electrode
SCD	Stray Current Dumpers
SCE	Saturated Calomel Electrode
SO^{2-}	Sulfate
TCE	Trichloroethylene
TDI	Toluene Diisocyanate
UN/ECE	United Nations Economic Commission for Europe
VAR	Vacuum Arc Re-Melting
VCM	Vinyl Chloride Monomer
VOCs	Volatile Organic Compounds

PREFACE

Despite commercial setbacks in recent years, the production of chlorine and alkalis still is of enormous importance, and chlorine and caustic soda both are among the 10 largest volume chemical products in the world. The number of final ends uses for the products of a chlor-alkali plant perhaps cannot be matched by any other single plant.

Chemistry lies at the heart of the chlor-alkali industry, driving the processes and innovations that enable the production and utilization of these essential chemicals. Understanding the fundamental principles and intricacies of the chemical reactions involved is key to unraveling the secrets of this industry and its impact on society.

This book comprises of eight-chapters; the first chapter provides an overview of the chlor-alkali industry and the processes involved in this industry. The second chapter deals with the significance and end uses of chlorine and caustic soda.

The third chapter provides an insight about the applied processes and technologies in the chlor-alkali industry and also discusses the processing, storage, and handling of chlorine. The fourth chapter discusses the anodes used in chlor-alkali industry along with the anode compositions, their physical characteristics and electrochemical behavior.

The fifth chapter describes the cathodes used in chlor-alkali industry along with their characteristics and morphology. The sixth chapter deals with the chemical engineering principles and practices adopted in the chlor-alkali industry. The seventh chapter discusses the issues for mercury cell technology in chlor-alkali industry and also highlights the new technologies such as membrane cell and diaphragm cell technologies. The eighth chapter provides an insight about the cost saving in chlorine plants and discusses the unique properties of titanium along with its uses in mercury, membrane, and diaphragm cell technologies.

The chemistry and chemical engineering in chlor-alkali industry is intended as a comprehensive resource for students, researchers, and professionals in the field of chemical engineering and industrial chemistry. It provides understanding of the chemical principles, processes, and applications within the chlor-alkali industry.

CHAPTER

1

Overview of the Chlor-Alkali Industry

Contents

1.1. INTRODUCTION

The worldwide capacity for chlorine production underwent a significant increase from around 22 million tons in 1970 to a staggering 53 million tons in 2002. It is projected to further escalate to 65 million tons by 2015 (O'Brien et al., 2005). The current chapter provides an overview of the principal manufacturing procedures and the variables that influence the expansion trajectory of the chlor-alkali sector.

Chlorine is commercially available in gaseous or liquid form, while caustic soda is marketed as a 50% or 73% solution or as dry beads or granules. A noteworthy market exists for caustic soda, ranging from 10–25% (Lakshmanan & Murugesan, 2014). The categorization of caustic soda products is based on the specific manufacturing technology employed, which includes the element mercury and the membrane cells or diaphragm. Most Chlorine (over 95%) and caustic (99.5%) substances worldwide are generated through cell technologies. Chemically synthesized soda ash contributes to approximately 0.5% of the total production of caustic soda, while potassium hydroxide (KOH) constitutes 2.5% of the global output of this highly alkaline substance (Crook & Mousavi, 2016).

This chapter provides an in-depth review of all three chlor-alkali technologies and a detailed discussion of the process specifics. The terminology and procedures delineated in this chapter ought to facilitate the reader in establishing a connection between the procedural stages and the domains of electrochemical and chemical engineering.

Did You Know?

Chlorine was first produced commercially using the chlor-alkali process in the early 19th century.

1.2. OVERALL PROCESS

1.2.1. Electrolyzers

The process of electrolyzing sodium chloride solutions results in the production of Chlorine and caustic soda (Morris, 1991). A significant proportion of the production cost of Chlorine is attributed to the electrical energy consumption necessary for the operation of the electrolytic cells. The efficiency of the electrolyzer is subject to variation, ranging from 85% to 98%, contingent upon the technology employed, the operational procedures of the electrolyzers, as well as the quality of the brine utilized in the procedure. Figure 1.1 displays the diagrams of the cell technologies.

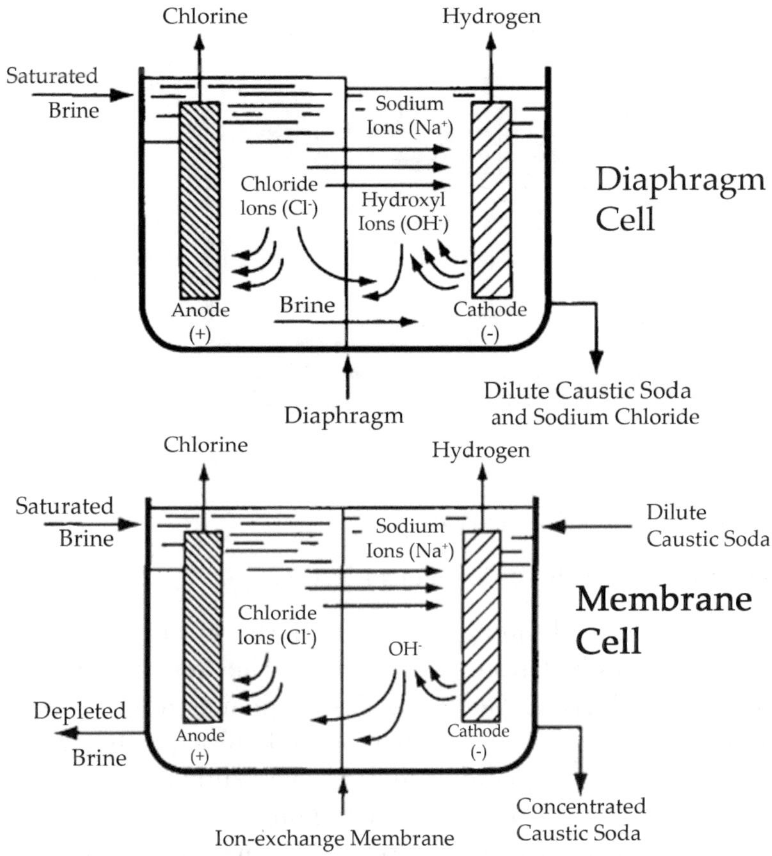

Figure 1.1. *Schematics of the diaphragm and membrane chlor-alkali cells.*

Source: Thomas F. O'Brien, Creative Commons License.

1.2.1.1. Mercury Cells

The mercury cell consists of a steel base, rubber-coated steel walls, and receptacles for the brine and mercury inflow and outflow, which are covered by an elastic rubber or rubber-lined steel lid (Campanari et al., 2019). The experimental setup involves using horizontally positioned, adaptable metallic anodes suspended from the upper portion of the apparatus. Meanwhile, the inclined base of the system facilitates the flow of mercury. The electric current is directed from the steel base toward the moving mercury (Figure 1.2).

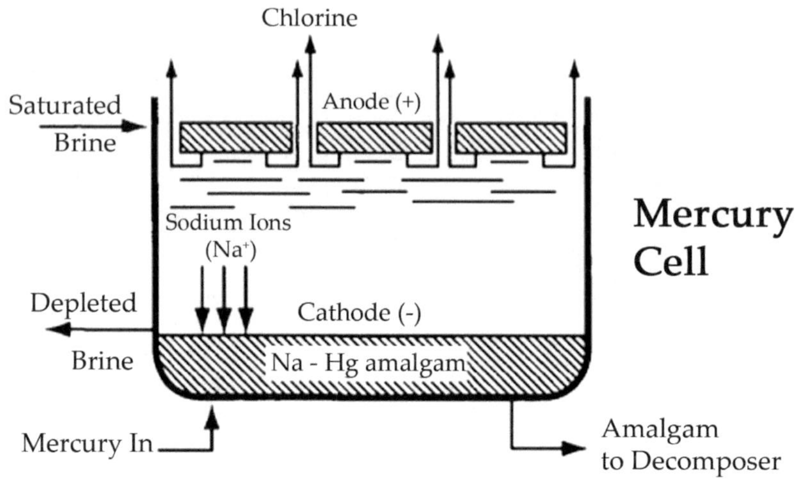

Figure 1.2. *Illustration of mercury cells.*

Source: Fumio Hine, Creative Commons License.

Chlorine is produced through the electrolysis of saturated brine at the anode, which is supplied from the end box. The trough's upper section carries the depleted brine. Upon reaction with mercury, the sodium or potassium cation undergoes amalgamation and subsequently flows into a vertical cylinder-like tank via the end box. The cell generates a potassium or sodium amalgam with a concentration ranging from 0.10 to 30% (Reis et al., 2009). When water is introduced to the decomposer containing graphite particles, the amalgam undergoes a reaction, forming hydrogen and either caustic soda or potash. Water vapor-saturated hydrogen is discharged from the upper portion, concurrent with the mercury fumes. The decomposer's concentration is either 48–50% of caustic soda or 45–50% of caustic potash. The brine that has been depleted is discharged from the terminal compartment. Certain cells are constructed to feature outlets for Chlorine and anolyte from the terminal box, which are segregated within the exhausted brine reservoir. The mercury derived via the decomposer is subsequently transported back to the cell.

1.2.1.2. Diaphragm Cells

The diaphragm or the membrane cell is a geometrically shaped container that features metallic anodes that are upheld from the lowermost section of the apparatus using copper base plates. The cathodes in the rectangular tank are vertical displays or punched platters that are interconnected from one side to the other. Asbestos dispersed in a slurry bath is utilized for vacuum deposition on the cathodes (Garcia-Herrero et al., 2017). The anode compartment is where saturated brine is introduced and subjected to electrolysis, resulting in the liberation of chlorine gas at the anode. The chlorine gas has reached a state of saturation with the water vapor at the vapor pressure associated with the anolyte. The transfer of the sodium ions from the anode chamber to the cathode section results in its reaction with a hydroxyl ion produced at the cathode throughout the hydrogen formation process from water molecules. This reaction leads to the formation of caustic soda. The diaphragm is a barrier to impede the retrograde movement of hydroxyl ions, which could otherwise undergo a chemical reaction with chlorine present in the anode section. The cathode compartment contains approximately 12% sodium hydroxide and 14% sodium chloride (Burney, 1993). The formation of sodium chlorate in the anolyte section is contingent upon the pH level of said compartment. Several non-asbestos separators have been successfully demonstrated in commercial settings, converting several asbestos-based cell circuits to these non-asbestos diaphragms.

1.2.1.3. Membrane Cells

Within a membrane cell, the cathode and anode compartments are segregated by separate ion-exchange membranes. Between the anode and the cathode is typically a bilayer membrane made of per-fluoro-sulfonic and per-fluoro-carboxylic acid-based films. The concentrated salt solution is added to the anode chamber, where the electrolysis process produces chlorine gas at the anode and aids sodium ions in migrating to the cathode chamber (Leddy, 1980). Only sodium ions and water molecules pass through the membrane, in contrast to diaphragm cells. The inert sodium chloride and other inert species are kept in the anolyte compartment. In the catholyte step, where the sodium ions interact with the hydroxyl ions produced during the process of forming hydrogen from water, 30–32% of caustic soda is added. The aforementioned method produces caustic, a highly corrosive chemical with a high concentration level ranging from 32 to 35%. The catholyte releases the hydrogen gas after it has completely absorbed water. The catholyte compartment is used to extract some of the caustic soda product, and the remaining substance is then diluted and reintroduced.

All three of the fundamental cell technologies used to produce chlorine result in the synthesis of hydrogen, sodium hydroxide, or KOH at the cathode and chlorine on the anode (Hou et al., 2018). The main point of distinction is to the technique used to avoid combining anolyte and catholyte.

In diaphragm cells, separation is attained through a porous separator, while in membrane cells, an ion-exchange membrane is utilized for this purpose (Lee et al., 2017). In the context of mercury cells, it is observed that the cathode functions as a separator owing to the formation of sodium amalgam. This sodium amalgam is then subjected to a reaction with water, producing sodium hydroxide.

1.2.2. Chlorine Processing

Chlorine gas comprises a byproduct of oxygen, moisture, and trace amounts of hydrogen. Furthermore, carbon dioxide, oxygen, and nitrogen infiltrate the equipment or pipes. Figure 1.3 depicts a basic diagram outlining the procedure for generating gaseous Chlorine or chlorine liquid from the cell gas (Margallo et al., 2016).

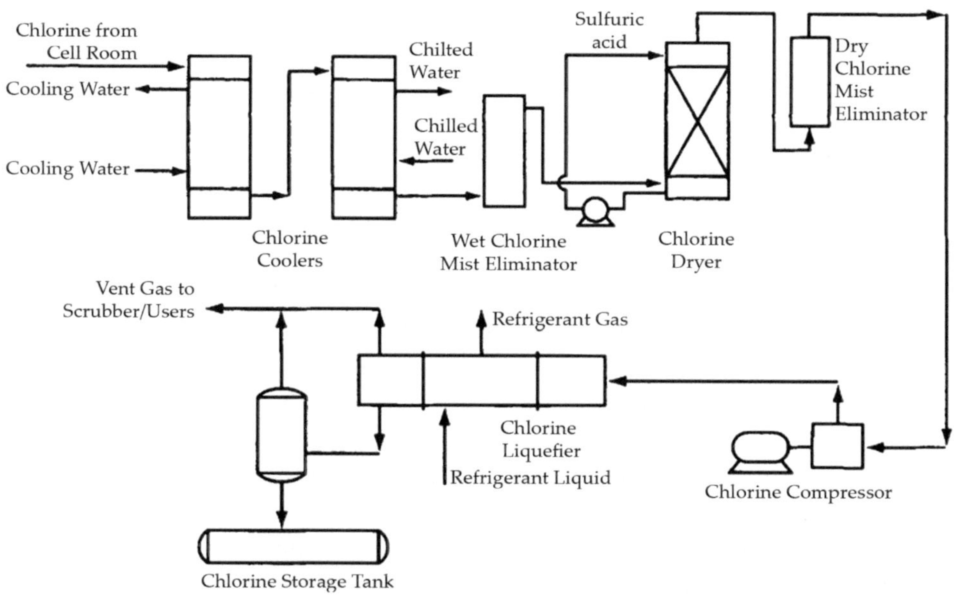

Figure 1.3. *Representation of process scheme for producing chlorine from cell gas.*

Source: Tilak V. Bommaraju, Creative Commons License.

The initial step in the process involves cooling Chlorine followed by its passage through demisters, which eliminate any water droplets and salt particulates. The cooled gas is directed towards a series of sulfuric acid circulatory towers for further processing. Typically, a triad of towers is employed for moisture elimination. The stream above frequently passes through desiccant demisters that remove moisture, subsequently proceeding to compressors (Bernaus et al., 2006). Subsequently, the gas in a compressed state undergoes a phase transition to a liquid state at a reduced temperature. Snift gas is in its non-condensed state and is utilized to produce hydrochloric acid or hypochlorite. If hydrochloric acid lacks a viable market, the snift gas is subjected to neutralization through caustic soda or lime, forming hypochlorite. Hypochlorite can be marketed as a bleaching agent or undergo decomposition to yield salt and oxygen.

1.2.3. Hydrogen Processing

Figure 1.4 illustrates the procedural stages entailed in the treatment of hydrogen derived from electrolytic cells (Karlsson & Cornell, 2016).

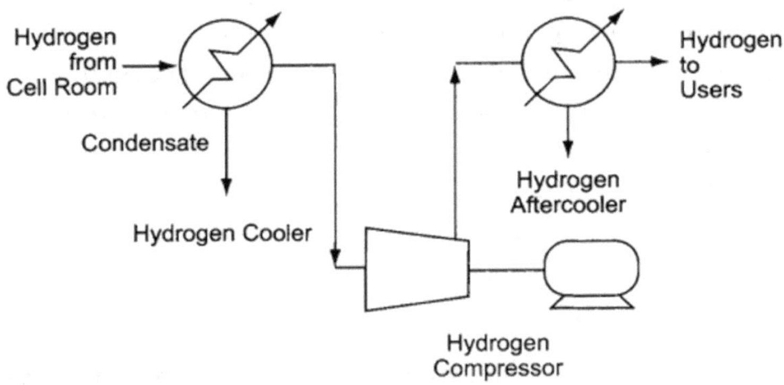

Figure 1.4. *Representation of process scheme for treating hydrogen from the electrolytic cells.*

Source: Robert, Creative Commons License.

Typically, the hydrogen gas from chloralkali cells is utilized to synthesize hydrochloric acid or as a fuel source for steam production. The hydrogen obtained inside the mercury cell undergoes a cooling process, after which the resulting condensed mercury is reintroduced into the cells (Hofmann et al., 2022). Typically, another treatment method is employed to eliminate the low concentrations of mercury present in hydrogen. Subsequently, the gas undergoes compression. When a customer requires hydrogen with minimal oxygen content, certain facilities employ a platinum catalyst to heat the hydrogen, thereby eliminating oxygen. The resultant hydrogen is cooled, compressed, and supplied to the customer via diaphragm and membrane cells. The recovery of heat values in the hydrogen vessel gas by cross-exchanging hydrogen vapor into the saltwater supplied to the cells.

1.2.4. Caustic Soda Process

Sodium hydroxide, commonly known as caustic soda, is commercially available in various forms, including 50% and 73% solutions and anhydrous flakes and beads. Mercury cells can generate caustic solutions with concentrations of 50% and 70% directly (Wang et al., 2014). Compared to caustic produced by membrane or diaphragm cells, this particular caustic exhibits higher purity. The caustic solution derived from the decomposer undergoes a cooling process and subsequently undergoes one or two cycles of filtration via an activated carbon medium. Upon undergoing the process of filtration, the concentration of mercury is reduced to parts per million (ppm) levels. Certain customers have adopted the use of membranes class caustic soda due to dissatisfaction with the current product. The caustic soda produced by the mercury cell method contains a few parts per million salt and less than five parts per million sodium chlorates (Roy et al., 2022). Assuming that trace amounts of mercury are deemed acceptable in the final application of caustic, it can be inferred that the caustic produced via the mercury cell process will possess the utmost level of purity achievable through electrolytic means.

Did You Know?

The term "caustic soda" comes from the latin word "causticus" meaning burning or corrosive.

The caustic membrane concentration is achieved through an evaporator, as depicted in Figure 1.5 (Zhang et al., 2021). The utilization of multiple-effect evaporators results in a significant improvement in the steam economy. The caustic solution produced by the membrane cell process typically contains 50% concentration. It has been observed that this solution contains approximately 30 parts per million (ppm) of salt and 5–10 ppm of sodium chlorate.

The catholyte solution obtained from the diaphragm cells comprises 129c caustic, 0.15–0.2% sodium chlorate, and 0.25–0.3% sodium sulfate. The catholyte undergoes evaporation, as depicted in Figure 1.6 (Wang et al., 2014). Upon reaching a caustic concentration of 50%, a significant portion of the salt undergoes separation, leaving a residual salt content of approximately 1% within the solution. The chlorate concentration in the 50% caustic solution is comparatively higher than that of the product obtained from membrane and mercury cells. The salt that is obtained through the process of evaporation and detached from the caustic is utilized for re-saturating the brine that is supplied to the cells. The production of 73% caustic involves the utilization of steam instead of water during the decomposition process, as well as the evaporation of half caustic. The production of anhydrous caustic is achieved by utilizing an expanding film evaporator, which operates under vacuum conditions at a temperature of 385°C (Moussallem et al., 2008). The alkali substance produced through the mercury-cell procedure is commonly called Rayon-grade caustic. The premium product known as "low salt" caustic soda encompasses rayon-grade caustic and membrane-grade caustic.

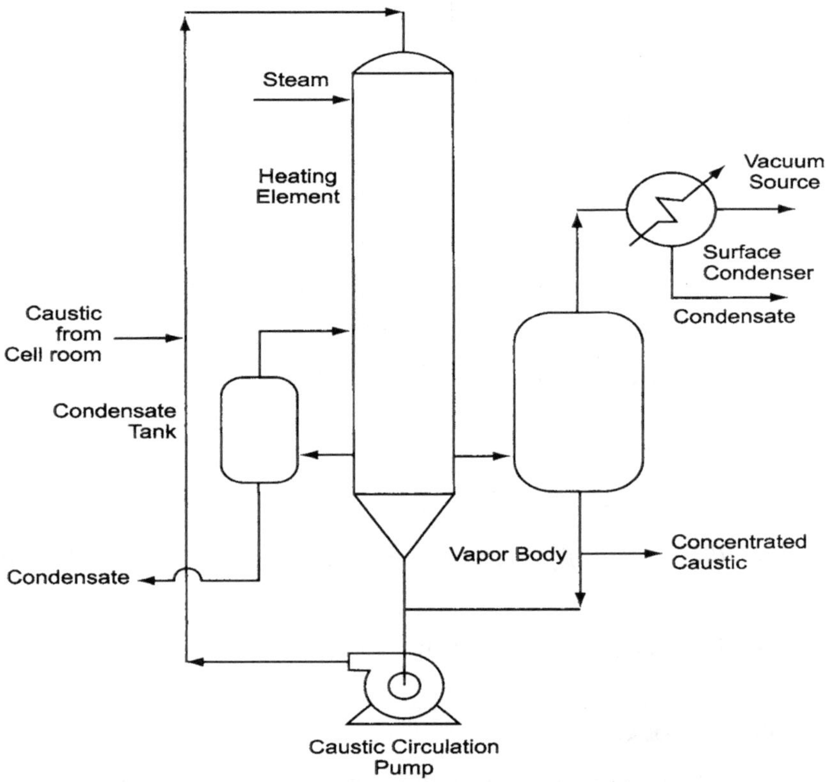

Figure 1.5. *Illustration of process scheme for concentrating membrane-cell caustic.*

Source: Steve Brien, Creative Commons License.

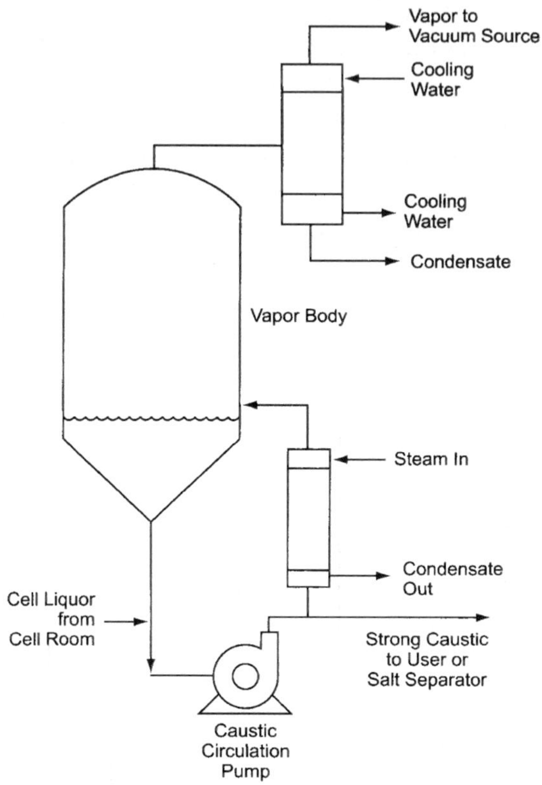

Figure 1.6. *Illustration of process scheme for concentrating diaphragm-cell caustic.*

Source: David Francis, Creative Commons License.

1.2.5. Brine Process

Sodium chloride is obtainable in a solid salt form, either through excavation or evaporating seawater. The aqueous solution of the substance above can be obtained through solution mining (Kinsey et al., 2004). The electrolytic cells' optimal current efficiency requires the elimination of impurities with varying concentrations present in the salt. The principal contaminants present are magnesium, calcium, and sulfur compounds. Minor impurities such as the element barium, manganese, strontium, aluminum, iron, vanadium, silica, chromium, titanium, and molybdenum may be considered undesirable depending on the specific chlor-alkali process employed.

The process of treating brine involves the utilization of a reactor in conjunction with the addition of sodium carbonate as well as caustic soda to facilitate the precipitation of carbonate of calcium and magnesium hydroxide from the brine, as demonstrated in the following equations (Lee et al., 2018):

$$Ca^{2+} + Na_2CO_3 \rightarrow CaCO_3 + 2Na+$$

$$Mg^{2+} + 2NaOH \rightarrow Mg(OH)_2 + 2Na+$$

The solid particles that have been formed are allowed to settle and subsequently extracted as a slurry. This slurry is then conveyed to a filter to separate and eliminate it as sludge. It is sometimes discarded with the remaining liquid effluent from the facility. The precipitate of calcium carbonate exhibits a significant weight and entrains the hydroxides of various elements such as aluminum, magnesium, and strontium (Khasawneh et al., 2019). The effluent from the settler, containing suspended solids ranging from 10 to 50 parts per million, undergoes filtration. The brine utilized in the mercury and diaphragm cell processes is deemed suitable for utilization and may be introduced into the electrolyzers. The diaphragm cell process involves heating the filtered brine and passing it through a salt saturator to enhance its concentration before introducing it to the electrolyzers. Acidification of the feed saltwater is a technique employed in certain plant species to enhance cell current efficiency. The process of acidification results in a decrease in alkalinity, thereby preventing the reaction between Chlorine and the anolyte compartment. This reaction would otherwise lead to the formation of hypochlorite and chlorate. The brine is acidified and subsequently introduced into the cells in the context of the mercury cell process.

The membrane cell process requires stricter brine specifications than the mercury and diaphragmatic processes, necessitating impurities at parts per billion (ppb) levels. The process of achieving this involves the utilization of a pre-coat type secondary filter to filter the brine, followed by applying an ion-exchange resin to eliminate impurities such as magnesium, calcium, the element barium, and iron (Garcia-Herrero et al., 2017). Aluminum removal through ion exchange is a viable option. The elimination of aluminum and silica from the brine discharged from the salt-based dissolver can be achieved by introducing magnesium chloride.

Remember:

The chlor-alkali process consumes a significant amount of energy. For the production of one ton of chlorine it requires around 3,500 to 4,000 kilowatt-hours of electricity.

1.3. ENVIRONMENTAL CONSIDERATIONS AFFECTING THE GROWTH OF THE CHLOR-ALKALI INDUSTRY

The chlor-alkali industry is frequently confronted with environmental challenges that can be categorized into two distinct groups: those that pertain to the process and technology and those that are determined by the end-use applications. This discourse pertains to issues related to the product.

Chlorine is utilized to produce both inorganic and organic compounds and indirect applications such as pulp and paper manufacturing and water treatment processes (Horowitz et al., 2014). Chlorine is known to possess acute toxicity towards humans, animals, and plants. However, it is noteworthy that Chlorine does not exhibit bioaccumulation in the human body and has no enduring impacts. Therefore, safety measures are implemented for transporting liquid Chlorine by carefully selecting appropriate containers, valves, and piping materials. The release of chlorine gas from containers due to leakage can result in the formation of a yellow cloud under high pressure, necessitating the evacuation of individuals in the surrounding area (Li et al., 2014). The statement above pertains to the Community Awareness and Emergency Response (CAER) initiative, a "Responsible Care" program component.

1.3.1. Organic Chemicals

1.3.1.1. EDC/VCM 7 PVC

Most global chlorine production is allocated toward fabricating polyvinyl chloride (PVC) and its associated intermediates. These plastic materials are extensively utilized in various applications, particularly in construction. During the 1990s, Western Europe emerged as the primary producer of PVC, with Asia and North America following suit.

The employment of PVC is currently a subject of debate among environmental agencies and PVC manufacturers, with four primary environmental concerns being the focal point of discussion. The potential environmental and health hazards associated with the incineration PVC include the generation of toxic dioxins, furans, and hydrochloric acid, both in the context of building fires and municipal waste incinerators (Bravo et al., 2016). The non-biodegradability of chlorinated plastic in garbage dumps is a significant concern. To date, no legal framework or regulatory measures impose limitations on the utilization of PVC. However, the utilization of the aforementioned is being criticized by environmental organizations.

Consequently, certain regional limitations have been enforced, particularly within the confines of Germany. Certain nations have prohibited manufacturing bottles utilizing PVC and PVC layers for packaging food within the packaging industry. HCl and dioxin emissions are deemed less critical due to the possibility of mitigating fugitive emissions through advanced waste incineration technologies (Muñoz & Navia, 2011). The utilization of stabilizers containing lead (Pb) and cadmium (Cd) has experienced a decrease in popularity, with stabilizers containing zinc (Zn) and calcium (Ca) being preferred instead. The utilization of plasticizers containing dioctyl phthalate in flexible PVC has been prohibited in toys intended for children. Phosphate esters have been utilized as a substitute for phthalates. It is noteworthy that recent research has revealed that the carcinogenicity of dioctyl phthalate is not as significant as previously assumed.

Vinyl chloride monomer (VCM), utilized in PVC manufacturing process, has been identified as a carcinogenic substance. As per the regulatory guidelines, the permissible threshold limit value for VCM in the air is 3–5 parts per million (ppm) (Schoenberger, 1990; Guandalini et al., 2017). Manufacturers of PVC have adhered to these limitations, resulting in effective management of the carcinogenic concerns associated with VCM in the final product.

1.3.1.2. Propylene Chlorohydrin

Propylene chlorohydrin is a crucial intermediate utilized in the manufacturing of PO, which serves as a fundamental component for producing urethane polyether polyols and propylene glycols. Propylene chlorohydrin is primarily produced in the USA and Western Europe, contributing to 74% of the global production output (Campanari et al., 2019). The primary ecological concerns pertain to the chlorinated garbage produced during the procedure and the management of the resultant calcium chloride sludge byproduct. The utilization of caustic soda can prevent the development of calcium chloride precipitate. The utilization of caustic is contingent upon the comparative costs of caustic and the expenses associated with the disposal of $CaCl_2$ sludge. A novel technique for the production of propylene oxide (PO) through the immediate oxidation of propylene has been developed by Olin (Song et al., 2018). There exist various peroxidation techniques that can be employed for the synthesis of PO. Potential advancements in non-propylene chlorohydrin-based pathways have the potential to decrease the necessity for Chlorine in the production of propylene oxide.

1.3.1.3. Chlorination Technologies

Chlorine finds application in producing numerous high-capacity commodities, such as chlorophenols, chlorobenzene, and chloroacetic acids. Hydrochloric acid, chlorosulfonic acid, thionyl chloride, phosphorus trichloride, phosphorus pentachloride, and phosphorous oxychloride are among the chlorinating agents utilized in manufacturing intermediates (Ullrich et al., 2007). One of the matters under consideration in the context of all of these chlorine reactions pertains to the generation of HCl gas, which may be alleviated by implementing exhaust gas containment mechanisms. Another environmental issue pertains to solid garbage and sewage pollution with both inorganic and organic contaminants, which emit extremely hazardous dioxins upon incineration. Although high-temperature burners can prevent this issue through complete thermal decomposition, the industry is concerned about public acceptance of this approach.

From the 1980s onwards, several chemical companies in Europe and Japan have implemented rigorous safety protocols that surpass official regulatory requirements. The allocation of resources towards ecological and safety strategies has resulted in a significant reduction in waste emissions.

1.3.1.4. Phosgene

Phosgene is a significant chlorine consumer that is an intermediary for generating isocyanates. It finds application in the manufacturing of various products such as polyurethanes, polycarbonates, pesticides, chloroformates, and carbonates, as well as in the production of pharmaceuticals. The global production of phosgene is divided between the USA and Western European countries, each contributing 41%, while Japan accounts for 12% of the overall production (Harris, 2008).

Phosgene is a substance of high toxicity, and its manufacturing, utilization, and transport are subject to legislative restrictions (Bravo et al., 2014). Manufacturing facilities that produce phosgene have implemented comprehensive safety measures that comply with the required standards. Nevertheless, the production of this substance may face limitations in densely populated neighborhoods or areas with stringent environmental regulations due to the influence of public risk perception.

1.3.1.5. Chlorine-Based Pesticides

Approximately 100 compounds based on Chlorine are utilized in hundreds of pesticides. Over 40% of these compounds exhibit insecticidal activity, 309c possess fungicide activity, and 15% exhibit herbicidal activity. These compounds are widely utilized in the agricultural sector. Table 1.1 displays the primary fungicides, insecticides, and herbicides utilized in certain global regions.

The necessity of these chemicals is contingent upon the particular market of significance within a given nation. Cotton, grains, fruits, and vegetables are significant commodities in the European market, whereas maize and soybeans are important in the US (Hutchison, 2008). The safeguarding of rice crops has become a pressing concern in both Hong Kong and Japan, where the employment of pesticides for preserving fruits and vegetables is also on the rise. Prominent agricultural markets in various nations encompass maize in the Republic of South Africa, grains and cotton fabric in Australia, fruits and grapevines in Chile, and cotton and rice in Colombia.

Table 1.1. Major Chlorine Containing Herbicides, Fungicides, and Insecticides with Regional Importance

	Major End Uses	United States	Western Europe	East Asia
A. Herbicide				
Acetochlor	Corn crop	No	–	–
Alachlor	Corn crop and soybeans	No	–	No
Atrazine	Corn crop	No	No	No
Bentazon	Soybeans crop	No	No	–
Chloramben	Soybeans crop	No	–	–
Chlorsulfuron	Wheat crop	No	–	–
Cyanazine	Cotton and corn crop	No	No	–
2,4-D	Cotton and corn crop	No	No	No
Dicyclofop-methyl	Wheat crop	No	–	–
MCPA	Wheat crop	No	No	No
Metalochlor	Corn crop	No	–	–

Metribuzin	Soybeans crop	No	No	–
Triallate	Wheat, maize, and oilseed	No	No	–
Trifluralin	Corn, soybeans, and cotton	No	No	No
B. Fungicide				
Anilazine	Non-agricultural	No	–	No
Captan	Vegetables, and non-agriculture	No	No	No
Carboxin	Peanuts crops	No	No	No
Chloroneb	Non-agriculture	–	–	–
Chlorothalonil	Vegetables, peanuts, and non-agriculture	No	No	No
Etridiazole	Non-agriculture	No	–	–
Iprodione	Rice, nuts, vegetables, and non-agriculture	No	No	No
Maneb	Fruits and vegetables	–	No	No
Metalaxyl	Nuts, tobacco, vegetables, and non-agriculture	No	–	–
Metiram	Potatoes crop	No	No	–
PCNB	Peanuts crop	No	No	No
Propiconazole	Small grain and non-agriculture	No	–	No
Thiram	Non-agriculture	No	–	No
Vinclozolin	Vegetables and non-agriculture	No	–	–
Ziram	Apples and stone fruit	No	–	–
C. Insecticide				
Carbofuran	Fruit, corn, nuts, beats, and maize	No	No	No
Chlorpyrifos	Corn crop	No	No	–
Cloethocarb	Corn crop	No	–	–
Dicofol	Cotton and citrus fruit	No	No	No

Dimethoate	Vegetables, fruit, and wheat	–	No	No
Endosulfan	Corn and fruit	No	No	No
Fonofos	Corn crop	No	–	–
Formetanate hydrochloride	Beet, vines, and fruit	No	No	–
Lindane	Corn crop	No	No	–
Methoxychlor	Non-agricultural	No	–	–
Thiodicarb	Cotton and soybeans	No	No	–

Source: Thomas F. O'Brien, Creative Commons License.

Although pesticides are effective in managing pest infestations and diseases, their usage can negatively impact the environment and plants through groundwater contamination. Additionally, the application of pesticides can lead to adverse reactions in humans. The circumstance above has resulted in implementing regulations by the Environmental Protection Agency (EPA) in the United States to restrict the utilization of pesticides that can potentially infiltrate the groundwater. Triazines, a class of compounds frequently found in pesticides, exhibit antibacterial and antiprotozoal properties (Heaton, 1993). Additionally, they are utilized in synthesizing atrazine, a weed management agent that has generated considerable debate.

A consensus has not been reached regarding the permissible upper limit of atrazine concentration. Various world health organizations have established the acceptable daily intake level for a particular substance at 2 μgL^{-1}, whereas the United States EPA has set a threshold at 3 μgL^{-1} (Cachada et al., 2009). Atrazine lacks regulation in Europe; however, the maximum allowable concentration for every single pesticide and all chemical pesticides combined has been established by the European communities' drinking water directive at 0.1 μgL^{-1} and 0.5 μgL^{-1}, respectively. Nevertheless, ongoing discussions and advocacy efforts persist in altering certain regulatory measures.

DDT, an insecticide utilized to control the typhus epidemic and malaria, was banned by the EPA in 1970 due to safety concerns. This ban resulted in the preservation of numerous lives. Several other insecticides, namely dielrin, aldrin, heptachlor, and chlordane, were prohibited due to their associated health risks. The US National Toxicological Program has classified 1, 4-Dichlorobenzene, toxaphene, and 1, 3-dichloroprene as suspected human carcinogens (Brinkmann et al., 2014). However, the EPA is conducting a special review of the latter two chemicals. The manufacture of herbicides containing 2,4,5-trichloropheonoxy acetic acid and fungicides containing 2,4,5-trichlorophenol has been terminated in numerous areas due to their inadequate biodegradability and tendency to accumulate in the food chain.

The production of chlorine-based herbicides has been affected by the environmental and health risks they pose. Furthermore, the consumption of insecticides containing chlorinated hydrocarbons has experienced a decline of over 95% since 1974. Even though this particular market segment represents a relatively small proportion of the overall chlorine consumption, it is improbable that the limitations imposed on Cl_2-based agricultural chemicals will significantly impact the chlorine industry. Nevertheless, the ongoing debates and discussions surrounding the use of pesticides have contributed to a negative perception of the chemical industry among the general public.

1.3.1.6. Chlorine Derivatives

Several chlorine derivatives present significant environmental hazards due to their potential to cause ozone depletion and contribute to the global warming phenomenon. The ozone depletion potential (ODP) is assessed by chlorofluorocarbon-11 (CFC-11), which is assigned a value of 1.0 (Han et al., 2014). Similarly, the global warming potential (GWP) is evaluated compared to carbon dioxide (CO_2), which is assigned a value of 1. The temporal scope of the GWP denotes the point at which the greenhouse effect ceases to exist. At the same time, atmospheric longevity pertains to the duration of decomposition for the specific species in question. Table 1.2 outlines the primary greenhouse emissions and their respective impact on the greenhouse effect (Melián-Martel et al., 2011).

Table 1.2. Representation of Greenhouse Effect

				Ozone	**CFC 11**	**CFC 12**
Concentration in the atmosphere (ppm)	354	1.72	0.31	0.03	0.00028	0.00048
Time in the atmosphere (yrs.)	120	10	150	0.1	60	30
Increment (to/yr.)	0.5	1	0.25	0.5	5	3

Relative global warming potential; = 1 (v/v)	1	21	206	2,000	12,400	15,800
Relative global warming potential; = 1 (w/w)	1	58	206	1,800	3,970	5,750
Contribution to the incremental greenhouse effect in the 80s (to)	50	13	5	7	5	12

Source: Fumio Hine, Creative Commons License.

The Montreal Protocol was established in 1987 by the United Nations to mitigate the depletion of the ozone layer and the release of greenhouse gasses in the atmosphere. Table 1.3 displays the CFC-phase strategies that were implemented in 1990 (Reig et al., 2014). The industrial sector has demonstrated a willingness to retrieve chlorofluorocarbons (CFCs) from outdated refrigeration equipment, adhering to the Montreal Protocol's regulations.

1.3.1.7. Chlorinated Solvents

The pertinent chlorinated solvents encompass all chloromethanes and chlorinated C_2 hydrocarbons, namely trichloroethylene, perchloroethylene (PCE), and 1, 1, 1-trichloroethane, which are extensively manufactured worldwide (Moorhouse, 2008). These products are subject to rigorous government regulations worldwide, and their production is expected to decline substantially in the coming decades.

Chlorinated solvents are commonly classified as air pollutants or toxic substances that may pose a risk to occupational health and safety. Therefore, the utilization of these substances has been greatly influenced by governmental policies and regulations during the previous quarter-century. The laboratory experiments on animals suggest that these solvents have the propensity to cause long-term impacts on human health.

The Montreal Protocol of 1990 imposed limitations on the manufacture and utilization of 1, 1, 1-TCA due to its substantial ODP, ultimately resulting in its discontinuation by 2005 (Li et al., 2021). The utilization of C_2 chlorine-based solvents for metal cleaning is a significant application. However, releasing these solvents into the environment is a pressing concern. The metal sector has implemented enhanced procedures to retrieve these solvents and mitigate atmospheric emissions. There is a current trend in the dry-cleaning industry to decrease the utilization of PCE. The process of phasing out PCE is currently in progress.

1.3.2. Inorganic Chemicals and Direct Application

1.3.2.1. Bromine

Although bromine is known to be highly toxic, it is noteworthy that compounds based on bromine are not considered to be toxic (Lakshmanan & Murugesan, 2014). However, it is worth noting that fire-suppression gasses, like halons, have been found to exhibit ODP. At the same time, lead anti-knocking has been identified as a soil and air pollution source. These substances have been subject to restrictions under the Montreal Protocol. Consequently, stringent regulations will be implemented on compounds containing bromine, particularly in industrialized nations.

1.3.2.2. Water Treatment

Approximately 5–6% of the global utilization of Chlorine is allocated towards the purification of drinkable, managed, and wastewater streams. The practice of disinfecting water for drinking with Chlorine, which has been used since the mid-1800s, has effectively eliminated fatalities caused by typhoid and numerous other contagious illnesses (Reis et al., 2009). Nonetheless, halogenated organic compounds, such as chloroform, are produced during chlorination.

Due to the carcinogenic properties of these species, regulatory bodies such as the US EPA and their international counterparts are actively pursuing measures to restrict the utilization of Chlorine in potable water. Adopting alternative disinfectant technologies by water treatment operations in the US, Western Europe, and Canada is expected to reduce chlorine consumption. The consumption of Chlorine is expected to rise in underdeveloped nations due to the installation of additional water treatment facilities (Du et al., 2022).

1.3.2.3. Pulp and Paper Industry

The primary concern associated with using chlorine in the paper and pulp sector is generating AOXs throughout the bleaching process involving chlorine or chlorine-based bleaching agents. The noxious adsorbable organic halides (AOXs) are discharged into the environment through gaseous emissions, aqueous effluents, and the processed, bleached pulp from the paper mill (Verma et al., 2022). Consequently, there has been a substantial reduction in the utilization of Chlorine for pulp bleaching. To mitigate the levels of AOXs present in the pulp, regulatory measures have been implemented in several regions, including the United States, Japan, Canada, and Europe. These measures have reduced AOX quantities from eight kilograms ton^{-1} of pulp to half a kilogram ton^{-1} of pulp.

1.4. HISTORY OF THE CHLOR-ALKALI INDUSTRY

In the latter half of the 1800s, Chlorine was primarily utilized in the textiles and paper sectors and was produced through the chemical reaction between manganese dioxide and hydrochloric acid (Wang et al., 2020):

$$MnO_2 + 4Cl \xrightarrow{100-110^\circ C} MnCl_2 + Cl_2 + 2H_2O$$

The enhancement of the general procedure economics was achieved by recycling manganese, leading to the terminology of the Weldon process. During the 1860s, the Deacon process was developed. This process involves the direct catalytic combustion of HCl with the air to generate Chlorine, as represented by the following equation:

$$4Cl + O_2(air) \xrightarrow{450-460°C;CuCl_2cat.} 2Cl_2 + 2H_2O$$

The necessary HCl for the reactions above was obtainable through soda ash production via the LeBlanc method (Ahmed & Foller, 2003).

$$H_2SO_4 + 2NaCl \rightarrow Na_2SO_4 + 2HCl$$

$$Na_2SO_4 + CaCO_3 + 2C \rightarrow Na_2CO_3 + CaS + 2CO_2$$

The employment of hydrochloric acid derived from the reaction above effectively mitigated the primary environmental concerns related to air and water pollution associated with the LeBlanc process while also facilitating the production of Chlorine. As of 1900, the Deacon and Weldon methods had yielded a sufficient quantity of Chlorine to produce approximately 1,50,000 tons of bleaching powder annually in England exclusively.

A significant revelation during this era was that steel exhibits immunity towards dry chlorine corrosion. In 1888, Badische Anilinund-Soda Fabrik was authorized to commence the initial commercial manufacturing and dissemination of dehydrated liquid Chlorine. The present methodology, which involves the utilization of H_2SO_4 for dehydration, subsequent compression of the gas, and its condensation via cooling, bears a striking resemblance to the technology under consideration (McCluney & Van Zee, 1989).

During the latter half of the 19th century, there was a shift towards utilizing the Solvay method to produce caustic soda, which gradually replaced the LeBlanc process. The resultant deficit of hydrochloric acid necessitated the exploration of an alternative pathway for the production of Chlorine.

While the initial discovery of Chlorine through saltwater electrolysis is credited to Cruikshank around the 1800s, it was not until 90 years later that the electrolytic technique was effectively employed for commercial purposes. Faraday established the principles that regulated the chemical breakdown of aqueous solutions in 1833. Subsequently, Cook & Watt were granted patents in 1851 and Stanley in 1853 for the electrolytic generation of Chlorine from brine.

1.4.1. Diaphragm-Cell Technology Development

The process of electrolyzing NaCl brine produces Chlorine at the end of the anode along with sodium hydroxide at the cathode (Dullien, 1958).

- At the Anode:

$$2Cl^- \rightarrow Cl_2 + 2e$$

- At the Cathode:

$$2H_2O + 2e \rightarrow H_2 + 2OH^-$$

- Overall Reaction:

$$2NaCl + 2H_2O \xrightarrow{2F} 2NaOH + Cl_2 + H_2$$

The primary challenge encountered during the electrolysis of salt solutions was the attainment of uninterrupted segregation between Chlorine, which was generated at the anode, and sodium hydroxide, which formed at the cathode (Olufemi et al., 2016). Completing hydrogen gasses and Chlorine within U-shaped tubes was a straightforward task. However, it was observed that the sodium hydroxide generated at the cathode underwent a reaction with Chlorine, leading to the formation of sodium hypochlorite. A British scientist, Charles Watt, developed the notion of a separator permeable to electric current. This separator permits the flow of electric current while maintaining the segregation of the cathode and anode products. The invention of the diaphragm cell dates back to 1851. The primary limitation of utilizing the Watt cell was its inadequate electrical generation potential. The advent of the dynamo in approximately 1865 facilitated the invention of electricity generators with adequate capacity and efficiency by Siemens, Edison, Wheatstone, Varley, and other notable inventors. This, in turn, made it possible to produce Chlorine and caustic soda through electrolysis viably.

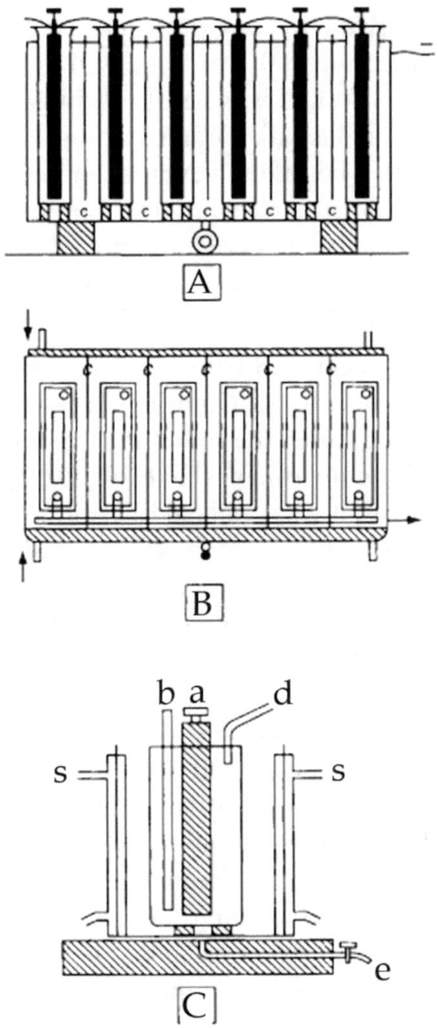

Figure 1.7. *Illustration of (A) and (B): Griesheim Elecktron cell bank; and (C) cross-section of Griesheim elecktron cell. a: anode; b: brine inlet; c: a cathode; d: brine outlet; e: caustic outlet; and s: steam.*

Source: David Francis, Creative Commons License.

Numerous nations made parallel advancements in diaphragm cells after the advancements in electricity generation. The Griesheim Company of Germany is credited with the development of the initial commercial cell for the production of Chlorine, which occurred in the year 1888 (Figure 1.7).

The United Alkali Company operated the Hargreaves-Bird cell, the initial diaphragm cell created in Great Britain in 1890. Every individual cell was comprised of a square iron enclosure coated with cement. The rectangular container measures 10 feet long, with a depth ranging from 4 to 5 feet and a width of 2 feet. It was partitioned into three compartments, each separated by two distinct diaphragms composed of asbestos sheets (Dibble & White, 1990). The experimental setup involved the placement of six anodes composed of carbon material within the anode compartment, while the cathode consisted of copper gauze affixed to the diaphragms. The fundamental technique of securing the diaphragm on the cathode persists in contemporary diaphragm cells.

The anode section was charged with brine that was fully saturated, and as a result of the electrolysis process, Chlorine was generated at the anode. In conjunction with NaCl and water, the sodium ions permeated the diaphragm and entered the cathodic compartment (Hu et al., 2012). The suppression of hydroxyl ion back migration was achieved by introducing CO_2 and vapor into the cathode compartment, forming sodium carbonate. The principal innovation of this particular cell was its upward diaphragm arrangement, which serves as the foundation for contemporary cells. Around 12 cells were operated at a current of 2 kA, resulting in an average current density of 200 A/m^2 at a voltage range of 4 to 4.5 V (Toma et al., 2019). During this process, approximately 60% of the salt underwent conversion to sodium carbonate. The Outhenin-Chalandre cell was employed in Italy and France and comprised an iron container partitioned into three sections. The outer compartments housed the cathode solution, while the inner one contained graphite anodes immersed in concentrated brine. The cell's partitioning walls were constructed by cementing cylindrical porcelain tubes that were unglazed and cylindrical to form the diaphragm (Stanley, 2008). The complexity of this cell surpassed that of the Griesheim cell and necessitated significant scrutiny.

In the United States, LeSueur established and managed a cell utilizing a percolating straight diaphragm in 1890 in the United States. The design above facilitated a continuous brine flow from the anolyte via the diaphragm, resulting in a superior efficiency to the non-percolating diaphragm cells prevalent in Great Britain and Germany during the same period. The percolating diaphragm cell developed by LeSueur, illustrated in Figure 1.9, serves as the foundation for all contemporary diaphragm chloralkali cells. This cell represents an enhanced iteration of the device depicted in Figure 1.8 (Collings et al., 1971).

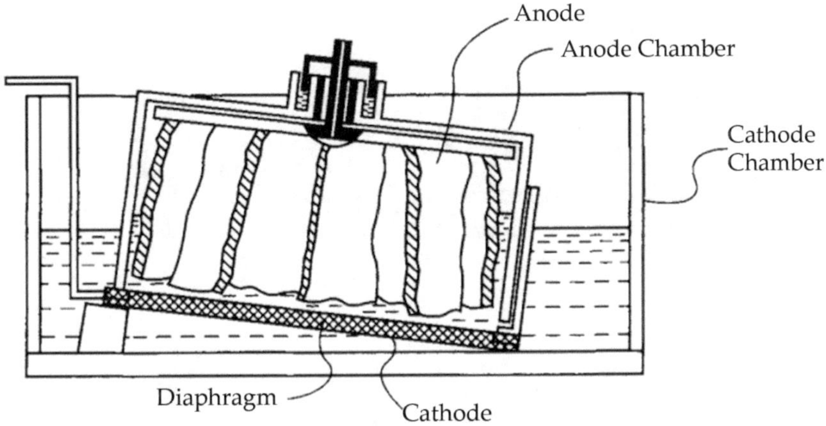

Figure 1.8. *Diagram of an original LeSueur cell.*

Source: Dr. James, Creative Commons License.

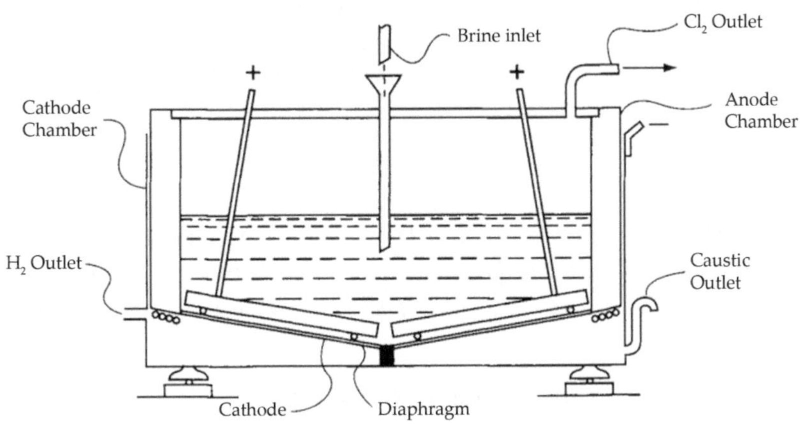

Figure 1.9. *Diagram of a modified LeSueur cell.*

Source: Thomas J. Navin, Creative Commons License.

The Billiter cell originated in Germany during the 1920s. The cellular configuration under consideration resembles the LeSueur cell, featuring a planar electrode and a diaphragm oriented parallel to the horizon in its initial blueprint. Modifying the cathode into a complicated shape increased current loads from 12 kA to 24 kA, as depicted in Figures 1.10(A) and (B) (Robinson et al., 1965). The diaphragm comprised a tightly knit steel mesh enveloped by a blend of extended barium sulfate and asbestos wool, forming a sealed structure. For the ensuing 80 years, asbestos served as the principal constituent of the diaphragm.

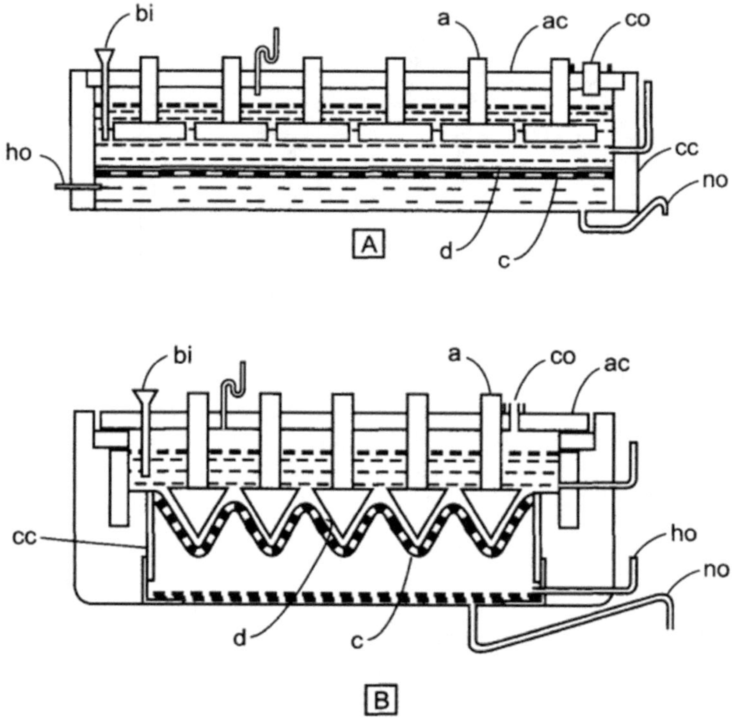

Figure 1.10. *Diagram of Billiter cell. (A) Flat cathode; (B) corrugated cathode. a: anode; c: a cathode; d: diaphragm; ac: anode chamber; cc: cathode chamber; bi: brine inlet; co: chlorine outlet; no: caustic outlet; and ho: hydrogen outlet.*

Source: Fumio Hine, Creative Commons License.

Between 1890 and 1910, there was notable progress in cell development. This was due to the introduction of LeSueur's and HargreavesBird's cell, which featured a vertical diaphragm, and the autonomous creation of synthetic graphite by Vaughn and Castner, and Acheson. Acheson also devised a technique to create synthetic graphite anodes. The cellular configurations were centered on robust concrete enclosures, typically of cylindrical or vertical shape, featuring graphite anodes that penetrate from the top and cathode plates secured to the side via bolts (Fell & Hutchison, 1971). The established fundamental cellular configurations encompassed rectangular upward cells, cylinder-like cells, and bipolar filter-press kind cells.

1.4.1.1. Vertical Diaphragm Cells

Among the several vertical diaphragm cells, the Townsend cell served as the precursor to contemporary diaphragm cells. The efficiency of the cell design utilized in the early 1900s needed to be improved due to Chlorine reacting with the back-migrating caustic. The Townsend methodology resolved the issue by applying a layer of kerosene to envelop the caustic (Asfour, 1983). The initial utilization of the Townsend cell was observed in Niagara Falls, where a plant consisting of 68 units of 2-kA cells was established. The primary output of the plant was Chlorine and caustic soda, with relatively low levels of hypochlorite and chlorate. During this particular era, a significant advancement occurred in which the saturated brine was fed through orifices to sustain the Intended brine level within the cell. This innovation resulted in the elimination of costly overflow controls.

In 1913, Clarence Marsh enhanced the Townsend cell's configuration by incorporating side-entry cathodes and anodes and introducing curved finger cathode plate electrodes. The design above served as a precursor to the cathode utilized in the Hooker Type S cell (Yamaguchi, 1990). The cathode was asbestos paper folded to conform to the combinations, with protective putty applied to the edges. Nonetheless, the disintegration of the cell putty resulted in the exposure of the cathode surface, leading to a decrease in the efficiency of the cell's current. In 1925, Stuart successfully resolved the issue of achieving full coverage of circular cathode pieces by utilizing a combination of asbestos fiber and diluted caustic liquor. This was accomplished by applying a vacuum, which facilitated the suction of the asbestos combination onto the cathode. Subsequently, the asbestos material adhered firmly to the cathode, perfectly fitting the cathode's shapes and curves. Asbestos vacuum deposition is regarded as a significant advancement in the chronicles of chloralkali cells. Figure 1.11 depicts the Marsh cell in both pre-and post-diaphragm deposition states (Cussler Jr & Dunlop, 1966).

After the innovation of placed asbestos diaphragms, Stuart created the Hooker "S" cell, featuring cathode fingers oriented vertically and equipped with deposited diaphragms. The "S" cells gained widespread popularity post-World War II and were gradually implemented globally. In 1960, approximately 60% of chlorine production in the US was attributed to using Hooker cells.

Hooker was responsible for the initial advancement of vertical diaphragmatic cells, whereas the Diamond Alkali Company devised an alternative cell configuration. Diamond obtained a license for a cell featuring cathode fingers that operate continuously, spanning the entire cell's width in 1959. The provision of support for the cathode hands at both extremities, in contrast to Hooker's cantilevered design that only provided support at one end, contributed to the maintenance of stability in the alignment of the anode and cathode (Arakawa et al., 2018). Furthermore, the orientation of the fingers is identical to the cell circuit, which allows for temperature expansion during operation and reduces the current path, leading to voltage savings.

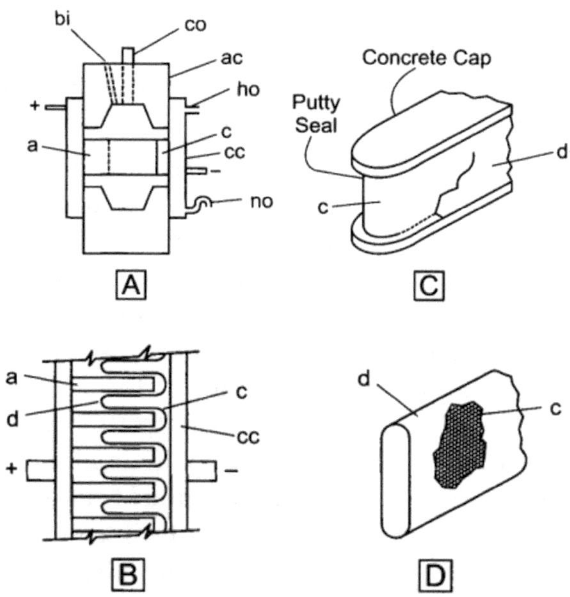

Figure 1.11. *Diagram of marsh cell. (A) Vertical section; (B) plan, sectional diagram; (C) original cathode with the asbestos paper diaphragm; and (D) Asbestos deposited cathode finger.*

Source: Tilak V. Bommaraju, Creative Commons License.

This characteristic has been implemented in nearly all contemporary cellular devices, including specific Hooker models delineated in Russian patents.

1.4.1.2. Developments in Anodes and Diaphragms

In the initial decades of the 20th century, the anode utilized for the production of Chlorine was limited to either magnetite or platinum. The prevalent utilization of graphite as an anode material persisted from 1913 until the mid-1970s due to the excessive expense of platinum and the existing current density constraint of 0.4 kA/m^2 associated with magnetite (Foller & Gu, 2016).

Acheson was responsible for developing artificial graphite anodes, which were subsequently enhanced in terms of their performance by Wheeler in 1919 through penetration with linseed oil (Laity & Miller, 1964). The regular substitute of anodes was deemed imperative as the gradual erosion of graphite blades led to an augmented cell voltage attributable to the amplified gap between the anode and cathode. The pursuit of dimensionally resistant anode structures as a substitute for graphite was instigated by this impetus. The utilization of activated titanium and platinum has been observed to provide exceptional corrosion resistance in the brine electrolysis service. Nevertheless, these anodes were plagued by either a limited lifespan or a substantial expense.

Henri Bernard Beer made significant contributions to the chloralkali industry by developing and submitting two patent applications in 1965 and 1967, which had a revolutionary impact. The patents above expounded on thermal decomposition methodologies for coating titanium substrates with combined crystals of platinum category metal oxides and valve metal oxides, with the valve-metal oxide proportion in these coatings exceeding 50% (Almeida Filho et al., 2011). The anodes in question were originally employed in DeNora mercury cells, demonstrating a protracted lifespan and a diminished cell voltage. In the subsequent years, most diaphragm-cell facilities worldwide underwent a conversion process to adopt the utilization of anodes branded as DSA (dimensionally stable anodes). Technology vendors have reconfigured their cellular arrangements to integrate said anodes to capitalize on the advantages of reduced energy consumption provided by metal anodes.

Following the creation of dimensionally resistant anodes, attention was redirected toward enhancing the quality of asbestos diaphragms. The substitution of fiberglass-reinforced nylon cell tops for concrete tops enhanced diaphragm longevity and voltage by removing the deteriorating byproducts of concrete. However, the diaphragms made of asbestos exhibited swelling during the operational period, leading to a rise in the cell's voltage.

Diamond Shamrock analyzed the behavior of asbestos diaphragms, and subsequently, a modified composition was formulated by blending asbestos with polytetrafluoroethylene (PTFE) (Florkiewicz, 2008). This resulted in creating a diaphragm that remained dimensionally stable upon heating. In 1972, commercial tests demonstrated an enhancement in cell voltage by implementing this particular diaphragm. This advancement facilitated the utilization of expandable anodes, which were concurrently developed.

One of the primary issues associated with prolonged contact with asbestos is its toxicological impact on the human body, leading to diseases such as asbestosis. This condition typically manifests 15 to 20 years following initial exposure. Consequently, the EPA has prohibited the utilization of asbestos in various applications such as carpeting felts, roofing felts, vinyl asbestos ground tile, and heat-resistant apparel. Despite the current exemption of the chlorine industry in the USA, which permits the usage of asbestos as diaphragmatic material, it is widely speculated that environmental demands will eventually compel the industry to transition towards non-asbestos diaphragm technology. Several nations have banned asbestos or established a preference for asbestos-free technology.

1.4.1.3. Diaphragm Cell Technologies

As previously stated, the Hooker "S" series cell was initiated by utilizing the vacuum-deposited diaphragm, developed in 1928 in conjunction with the upward finger anode-cathode design. The cell type denoted by the letter "S" persisted until 1966, when it was superseded by the singular C-60 cell. The year 1970 marked the introduction of the "H" series cell, which featured metal anodes replacing the previously utilized graphite anodes. Table 1.3 presents a comprehensive overview of the primary advancements in Hooker diaphragm-cell design (Thiel et al., 2017).

Table 1.3. Summary of Major Changes in Hooker Diaphragm Technology

Date	Cell	Key Innovation
1928–1948	S-1	First diaphragm cell is deposited. Minor electrical circuit adjustments consistently improve permissible amperage.
1948	S-3	Doubled S-1 anode and cathode regions to double capacity.
1951	S-3A	The greater current density required larger conductors.

1956	S-3B	More C and H_2 separating area for chlorine synthesis.
1960	S-3C	Economically optimized conductors.
1963	S-4	Heightened the anode and cathode and added 5 electrode rows.
1964 S-	S-4B	Cost reduction by lowering grid height. Disengaging space expanded.
1966	C-60	Redesigned cathode to decrease copper without boosting voltage.
1970	H-1	S-4B metal anodes increased the density of currents and cell life.
1970	H-2, 2A	Improved cathode and foundation copper for higher electrical currents using basic S-4 and C-60 models.
1971	HC-4B	S-3C metal anodes increased current density along with cell life.
1972	H-4	Improved cathode and anode area for capacity.
1978	H-4M	Polymer added to asbestos creates a durable, energy-efficient diaphragm.
1979	H-4/84 (or LCD)	By lowering the anode-cathode gap and adding fingers to the H-4 cathode shell, the LCD cell had a 12% lower current density.

Source: Tilak V. Bommaraju, Creative Commons License.

In 1959, Diamond Alkali commenced licensing technology that bore a striking resemblance to the Hooker cells. The primary cellular unit in the period spanning from the late 1950s to the mid-1960s was the D-3 cell, which functioned at a current intensity of 33 kA (Prout & Moorhouse, 2012). As previously mentioned, the D-3 model featured cathode fingers firmly affixed on either side of the cathode assembly. The cathode assembly consisted of a rectangular steel container containing an asbestos diaphragm that had been deposited onto it. The cell was finalized by adding a cathode assembly with a solid head.

Subsequent iterations of Diamond cells encompassed the DS-45, DS-31, and DS-85 variants, which were equipped with metallic anodes. The DS-85 variant facilitates a maximum amperage of 150 kiloamperes. The cell "bottom" component incorporates a patented configuration comprising a welded steel framework that sustains a level copper grid for conveying electrical current to the anodes (Bommaraju et al., 2007). The copper is safeguarded against the corrosive effects of the anolyte by utilizing a mat that is resistant to corrosion. The metal anodes, which are arranged in a vertical orientation, are affixed to the base through bolts that pass through the mat. Like Hooker designs, the exterior or cell head is composed of a distinctive polyester resin rather than concrete (Huber & Benzing, 2005). Implementing this particular cover effectively mitigates anolyte contamination that may arise from concrete degradation, a problem previously encountered in similar cells. Consequently, the incidence of diaphragm blockage is significantly reduced.

At present, Eltech System Corporation is engaged in the promotion of Hooker cells along with Diamond cells. Additional diaphragm-cell technologies were developed in the 1970s.

1.4.1.4. Bipolar Diaphragm Cells

The development of the filter-press cell for the production of Chlorine and caustic was the result of the efforts of multiple pioneering researchers. Dow played a pioneering role in the advancement of bipolar filter-press cells, which spanned multiple years, owing to their numerous benefits, such as reduced interconnection between cells and floor space, as well as a low initial investment. Dow's initial patents were granted in 1911 or 1913 (Ursua et al., 2011). The filter-press cell configuration granted a patent in 1921, comprises concrete filter-press modules that feature graphite rods that are embedded and protrude into the anode section and the cathode screen on the opposite side.

The US patent 2282058, authored by Hunter-Otis and Blue, presents an enhanced bipolar design of Dow (Satoh et al., 2014). This design involves casting a graphite-dividing wall into a concrete filter-press frame and incorporating graphite anodes that project upwards from the backing plate. The cathode hands were constructed using a steel wire screen enveloped with asbestos cardboard and secured onto a screen supporting plate located in the neighboring unit cell. Lucas and Armstrong implemented enhancements to the cell design by utilizing bipolar units that fit snugly into grooves within the cell box, thereby mitigating issues related to sealing. The electrode unit consists of cathode fingers made of a stainless-steel screen and graphite anodes that project upwards from a steel foundation plate, forming an integrated structure. The Glanor electrolyzer, a bipolar filter-press diaphragmatic cell, was collaboratively developed by PPG and

DeNora. The primary design characteristic entails the utilization of a bipolar electrode, wherein one facet functions as an anode while the other serves as a cathode (Suleman et al., 2016). The electrode consists of a steel plate with anode fingers on one end and cathode fingers on the opposite side.

The bipolar element comprises a channel frame that encircles the anodes and houses the electrode (Snyder et al., 2003). The anolyte environment is known to cause corrosion. Hence the titanium lining is incorporated to provide protection. The process of assembling a cell involves clamping together components, with the anode end of one element confronting the cathode side of a different one through tie-rods. The resultant configuration bears semblance to the plate and frame filter press, a frequently employed apparatus for filtration procedures. Therefore, the expression "bipolar filter-press cell" is utilized.

1.4.2. Mercury Cell Technology

A distinct approach to the generation of caustic soda and Chlorine entails the utilization of mercury cells, which exhibit dissimilar cathode reactions compared to those in diaphragm cells (Hławiczka & Fudała, 2008).

The mercury cell is composed of two fundamental components, namely its electrolyzer and its decomposer. The primary electrochemical reactions can be delineated as follows (Sznopek & Goonan, 2000):

- In the Electrolyzer:
- At the Anode:

$$2Cl^- \rightarrow Cl_2 + 2e$$

- At the Cathode

$$2Na^+ + 2Hg + 2e \rightarrow 2Na - Hg$$

- Overall Reaction:

$$2NaCl + 2Hg \xrightarrow{2F} 2Na - Hg + Cl_2$$

- In the Decomposer:
- At the Anode:

$$2Na - Hg \rightarrow 2Na^+ + 2Hg + 2e$$

- At the Cathode:

$$2H_2O + 2e \rightarrow 2OH^- + H_2$$

- Overall Reaction:

$$2Na - Hg + H_2O \xrightarrow{2F} 2NaOH + H_2 + 2Hg$$

The general procedure entails the movement of a refined and concentrated saline solution through a prolonged, marginally slanted channel between a superficial stream of mercury that flows in the same direction and an arrangement of electrodes (constructed from graphite or metal). The lower facets of these electrodes are in proximity to the surface of the mercury and are parallel to it. A gas-tight cover is affixed to the cell trough, which functions as a support for the anodes. In the process of electrolysis, the anodes facilitate the liberation of Chlorine (Pirrone et al., 2010). In contrast, the outermost layer of the mercury cathode enables the discharge of sodium ions, resulting in a sodium amalgam with a low concentration of sodium (0.1–0.3% by weight). Consequently, the reactivity of sodium ions with the water-based electrolyte to produce caustic soda or dispersed Chlorine to regenerate sodium chloride is significantly impeded. Upon reaction to water in the decomposer, the amalgam undergoes a chemical transformation, producing highly concentrated caustic of pure nature.

The process of producing Chlorine and caustic using a mercury cell was pioneered by Hamilton Y. Castner and Karl Kellner in 1892. It is worth noting that the first patent related to the mercury cell was British patent 4349, which was granted to Nolf in 1882. After engaging in patent conflicts, Castner & Kellner collaborated and transferred their collective patent ownership to Solvay et Cie. The initial Castner mercury cell consisted of a fixed, sturdy cast-iron foundation that contained drilled cylinders for the ram pumps utilized to facilitate the circulation of the amalgam (Tonini et al., 2003). Two distinct anode compartments facilitated the electrolysis process, while an intermediate compartment was utilized for amalgam decomposition, ultimately producing caustic production. The ram pumps of this cell experienced seizure due to the formation of caustic soda deposited.

In tandem with the progress above in the US, Brichaux & Wilsing of Solvay constructed the inaugural "long" cell in Germany in 1898 (Masding & Browning, 2008). This particular cell exhibited contemporary features of compelled mercury circulation.

The proliferation of synthetic fibers, exemplified by rayon, has significantly increased the requirement for high-grade (devoid of salt) caustic soda (Esbri et al., 2009). During the period surrounding World War II, a considerable quantity of mercury-cell factories was constructed.

Following 1945, there was a gradual increase in cell capacities from approximately 10 kA to 200–300 kA. The technology in question is supplied by various entities such as Hoechst-Uhde, De Nora, Krebs, BASF, Krebskosmo, Kureha, Solvay, Toyo Soda, Olin-Mathieson, and Asahi Glass (Walker, 2016). The mercury cell's enhanced capability was contingent upon using proficient silicon rectifiers, which superseded the antiquated mercury-arc rectifiers. The divergent characteristics of these mercury cell technologies are attributed to various design elements, such as the mode of mercury delivery to the electrolyzer, decomposer configuration, anode and cathode support structure, cell cover types, and mercury inventory prerequisites, all aimed at enhancing economic viability.

The diaphragm-cell process is the predominant technology employed globally and in the United States for producing Chlorine and caustic. This method accounted for 37% of the worldwide chlorine output. The diaphragm cell technology accounted for 679c of chlorine production in the United States (Mihaiescu et al., 2012). Initially, the technology utilizing mercury-cells held a dominant position. After the incidents of mercury poisoning in Minamata and Niigata, Japan, in 1972, there has been a decreasing inclination towards employing this technique to produce Chlorine. Japan was the pioneer nation in transitioning from mercury-cell technology, and presently, there are no operational mercury-cell facilities in Japan. Mercury-cell technology is a prevalent practice in Western Europe. In 2001, the global share of mercury-cell technology amounted to 18%. Mercury cells produce 11% of the total chlorine output in the US. It is improbable that this circumstance will significantly alter in the immediate time frame. The present discourse delves into the ecological problems and concerns associated with the mercury-cell process.

1.4.3. Membrane-Cell Technology

The membrane cell is a viable substitute for both diaphragm and mercury cells. The membrane cell is a variant of the diaphragm cell, in which the diaphragm is substituted with a selectively permeable ion-exchange membrane (Nidola, 1986). The selectively permeable membrane restricts the translocation of anionic chloride species while facilitating the movement of cationic sodium species across the membrane. The process of membrane-cell involves introducing brine containing either potassium or sodium chloride into the cell, which is then uniformly distributed among the anode compartments (Mo & Ng, 2010). Simultaneously, water is supplied to a separate header and directed toward the cathode compartments or an externally recirculating catholyte stream. The electrode reactions in this context exhibit similarity to those observed in a diaphragm cell. The anolyte, the depleted brine, and the catholyte, the caustic soda liquor, are separated and transported through distinct headers. The anolyte carries chlorine gas, while the catholyte carries hydrogen gas.

Multiple companies were involved in the advancement of membrane cell technology. The investigation of particular perm membranes at a conceptual level was initiated during the mid-1940s. From the middle of the 1950s through the early 1960s, diamond and ionics collaborated to advance the development of the membrane cell (Kim et al., 2009). The present study exhibited the practicability of the notion; however, it encountered significant issues such as carbon anode erosion, elevated cell voltage, and membrane lifespan that was exceedingly brief. The project remained inactive for approximately a decade.

During the latter half of the 1960s, the apprehension surrounding environmental issues linked to mercury-cell plants prompted a renewed fascination with membrane-cell technology. During the early 1970s, the initial challenges related to membrane cells were gradually mitigated. Utilizing metal anodes has effectively addressed issues related to the erosion of graphite anodes. During fuel cell research, DuPont created a perfluorinated ion-exchange membrane as a byproduct of their involvement in the space program. The membrane, originally called "XR," was subsequently registered under the trademark "Nafion" (Batko et al., 2020). These membranes exhibited favorable ion-exchange characteristics and demonstrated remarkable durability in the challenging conditions of the chlor-alkali cell.

Following DuPont's introduction of Nafion membranes, Diamond Shamrock increased its research endeavors in the field of membrane-cell technology. Preliminary investigations involving the membranes were conducted in 1970 utilizing laboratory cells. A pilot plant and electrolyzer of commercial scale were operationalized in Painesville, Ohio, in 1972 (Gai et al., 2016). After the initial development, a pilot facility capable of processing 20 tons per day was constructed in Muscle Shoals, Alabama, four years later. The integration of this unit was carried out with an extant mercury-cell plant.

The initial commercial membrane cell introduced by Diamond was the "DM-14 Electrolyzer," a monopolar cell. Several distinct membranes are placed within frames of anode and cathode that are fastened together in a configuration resembling a filter-press (Li & Li, 1993). The active electrode area of each electrolyzer unit, consisting of an anode, membrane, and cathode, is 1.41 m^2. The current concentration of 3.1 kiloamperes per square meter produces 0.14 metric tons of caustic daily. The operational current of the electrolyzer is contingent upon the number of unit cells that are fastened together, with a range spanning from 5 kA to more than 150 kA. The potential of the electrolyzer remains unaffected by the number of membranes installed due to the monopolar configuration of the cells. The DM-14 was observed to function at a voltage of 3.9 V while maintaining a density of current of 3.1 kA/m^2.

Like Diamond, Hooker Chemicals was keenly inclined towards ion-exchange membranes dating back to the mid-1940s. The first membrane cell was patented by Osborne et al. from Hooker Chemical Corporation. The filing date of this patent is January 23, 1952 (Knepper et al., 2011). Hooker initiated a significant developmental program until 1957, after obtaining this patent. After a period of low activity, the "Nafion" membranes gained attention in the field of membrane-cell technology.

The culmination of research efforts in the early 1970s resulted in a license of Hooker Chemicals' MX cell to Reed Paper Ltd. (Selvaraj et al., 2016). Subsequently, a membrane plant with a capacity of 45 tons per day was constructed in Dryden, Ontario. The membrane above cells was implemented to replace 13-year-old mercury cells, which were gradually phased out due to mounting apprehension regarding mercury pollution issues in Canada. The operational commencement of the plant occurred on the 19th of November 1975.

During a concurrent timeframe, Asahi Chemical initiated an experimental operation utilizing ion-exchange membrane cells with the implementation of DuPont's Nafion. The year 1975 marked the commissioning of a commercial plant in Nobeoka, Japan, which could produce approximately 10 tons of caustic per day.

Subsequently, several membrane-based cell technologies were devised in Japan, offering a chlor-alkali process devoid of pollution. The Japanese have made significant contributions in the field of electrolysis, including the development of composite membranes and various designs of electrolyzers. Japan was the pioneering nation in adopting membrane cell technology for chlorine production on a large scale. As of January 2003, membrane-cell technology accounted for 35% of global chlorine production, producing approximately 52,000 metric tons of caustic per day.

Numerous entities involved in chlor-alkali production and technology licensing have enhanced and perfected membrane cell configurations to achieve energy efficiency and durability. The novel cellular configurations integrated the zero-gap principle, eradicating the electrolyte interstice amidst the electrodes and the membrane (Jacob et al., 1991).

Two distinct categories of electrolyzers have been devised: mono-polar and bipolar. The mono-polar configuration arranges all anode and cathode elements in parallel, resulting in an electrolyzer operating at low voltage and high amperage. The bipolar configuration involves interconnecting the cathode of one cell with the anode of an adjacent cell, resulting in a series of cell arrangements (Mo & Ng, 2010). This arrangement yields an electrolyzer with high voltage and low amperage. This study examines the benefits and drawbacks of the various electrolyzer configurations.

An additional noteworthy advancement pertains to the retrofitting of diaphragms through the utilization of ion-exchange membranes, which serve as a substitute for asbestos diaphragms within a pre-existing cell. In retrofitting, the membranes are molded and thermally fused to envelop the electrodes, effectively substituting asbestos (Dias, 2010). The approach was implemented by Chlorine Engineers for DS cells and by Kanegafuchi Chemical for H-series cells. Various companies, such as Tokuyama Soda, Mitsui Chemical, and Toyo Soda, have demonstrated the effectiveness of the diaphragm retrofit technology.

Initially, the membranes fabricated using perfluorosulfonate polymer Nation were operationally feasible only under low caustic concentrations, owing to the substantial reduction in caustic efficiency observed at high caustic strengths (Lei et al., 2018). As a solution to this issue, researchers developed asymmetric membranes that feature sulfonic acid groups on the anode end and transformed groups on the cathode side. Subsequently, a membrane composed of perfluorocarboxylate, known as Flemion, was developed, demonstrating superior durability against the adverse effects of caustic back-migration. The advantageous properties of persulfonate based membranes, namely their low electrical resistivity, and perfluorocarboxylate membranes, which exhibit low caustic back-migration characteristics, were combined through the conversion of the sulfonic acid groups of Nafion to groups of carboxylates on the cathode side. Contemporary membranes comprise a composite structure with a PTFE reinforcing fabric, a perfluorosulfonate polymer layer, and a perfluorocarboxylate polymer. These components are firmly joined together to form a cohesive unit.

1.4.4. Caustic Soda

During the early 1900s, caustic soda production was carried out through the lime-soda process, despite the availability of caustic soda via the chlor-alkali electrolysis process. This was due to the limited demand for Chlorine at the time. The demand for Chlorine experienced a significant surge during the First World War, owing to its requirement for synthesizing diverse chemical compounds (Morris, 1990). Consequently, there was a rise in the market share of caustic obtained through the electrolytic process. The Second World War had a comparable impact. Currently, most caustic soda is produced as a byproduct of electrolytic chlorine production.

> **Did You Know?**
>
> Chlorine and caustic soda are used extensively to disinfect drinking water, wastewater, and swimming pools, helping to prevent the spread of waterborne diseases.

1.4.4.1. Lime Soda Process

This involves the reaction of lime and sodium carbonate (Xia et al., 2022):

$$Ca(OH)_2 + Na_2CO_3 \rightarrow CaCO_3 + 2NaOH$$

Calcium carbonate precipitates from the solution of calcium hydroxide due to its comparatively lower dissolution than the latter. The temperature range of 80 to 90°C is optimal for the reaction. Calcium carbonate undergoes sedimentation, while the excess liquid is a dilute caustic soda solution. The solution is subsequently concentrated through the process of evaporation.

1.4.4.2. Electrolytic Process

The production of caustic soda is a byproduct of the electrolytic process of sodium chloride brine, which involves diaphragmatic cells, mercury cells, and membrane cells and is produced in conjunction with Chlorine.

The diaphragm cell can generate a caustic soda solution comprising 11% caustic soda and 15% salt while exhibiting a reduced level of sodium chlorate concentration (von Weymarn et al., 2002). The remedy above, commonly called cell liquor, is subjected to evaporation procedures resulting in caustic soda production with a concentration of 50%. As evaporation occurs, the concentration of caustic substances increases, leading to salt crystallization. The 50% caustic soda solution contains approximately 1% salt. The mercury cell can generate caustic soda at a concentration of up to 70% and produce 50% caustic soda. The substance in question exhibits a high purity level, containing only a few parts per million of salt and fewer than five parts per million of sodium chlorate. Notwithstanding, the caustic soda derived from mercury cells contains minimal mercury, thereby rendering it undesirable to a significant number of consumers.

The membrane cell process yields approximately 35% of caustic soda, which is subsequently concentrated to 50% through evaporation. The utilization of membrane cell technology for producing caustic soda is deemed more favorable compared to the diaphragmatic and mercury-cell methods.

1.5. FUTURE DEVELOPMENTS

During the past century, considerable research has been conducted on chlor-alkali technologies to achieve cost-effective and efficient production of Chlorine and caustic soda while adhering to all governmental regulations for environmental safety. The energy consumption rate has significantly declined, from approximately 4,000 AC kWh/ton of caustic in the 1950s to roughly 2,500 AC kWh in 1998 (Rosales-Huamani et al., 2021). This decrease can be attributed to the emergence of dimensionally stable anodes and the implementation of optimal cell design and operation practices. Further enhancements, such as integrating fuel cell principles and advancements in electrocatalyst design, are poised to yield additional reductions in energy consumption and, by extension, the expenses associated with the manufacture of Chlorine and caustic.

1.6. SUMMARY

The chlor-alkali industry is an essential sector involved in the production of chlorine, sodium hydroxide, and hydrogen gas through the electrolysis of salt solutions. These chemicals have widespread applications in various industries and are fundamental to the production of numerous consumer and industrial products. The chlor-alkali process involves the electrolysis of a brine solution, typically consisting of sodium chloride dissolved in water. The process take place in specialized electrolytic cells called chlor-alkali cells, which are often made of steel or other suitable materials. In the electrolytic process, an electric current is passed through the brine solution, causing the salt to split into its constituent ions. The chlor-alkali industry has made significant strides in improving environmental performance. The transition from mercury-based cell technology to more environmentally friendly alternatives such as membrane and diaphragm cells has greatly reduced the release of mercury into the environment.

REFERENCES

1. Ahmed, S., & Foller, P., (2003). Responding to a harsh business environment: A new diaphragm for the chlor-alkali industry. *The Electrochemical Society Interface, 12*(4), 32–34.
2. Almeida Filho, E. M., Vilar, E. O., & Feitoza, A. C. O., (2011). Physical–chemical characterization and statistical modeling applied in a chlor-alkali diaphragm-cell process. *Chemical Engineering Research and Design, 89*(4), 491–498.
3. Arakawa, T., Xie, R., Seshima, F., Toma, K., & Mitsubayashi, K., (2018). Air bio-battery with a gas/liquid porous diaphragm cell for medical and health care devices. *Biosensors and Bioelectronics, 103*(1), 171–175.
4. Asfour, A. F. A., (1983). Improved and simplified diaphragm cell design and analysis technique for calibration. *Review of Scientific Instruments, 54*(10), 1394–1396.
5. Batko, K. M., Ślęzak, A., & Bajdur, W. M., (2020). The role of gravity in the evolution of the concentration field in the electrochemical membrane cell. *Entropy, 22*(6), 670–680.
6. Bernaus, A., Gaona, X., van Ree, D., & Valiente, M., (2006). Determination of mercury in polluted soils surrounding a chlor-alkali plant: Direct speciation by X-ray absorption spectroscopy techniques and preliminary geochemical characterization of the area. *Analytica Chimica Acta, 565*(1), 73–80.
7. Bommaraju, T. V., Orosz, P. J., & Sokol, E. A., (2007). Brine electrolysis. *Electrochemistry Encyclopedia*, *1*, 1–17.
8. Bravo, A. G., Cosio, C., Amouroux, D., Zopfi, J., Chevalley, P. A., Spangenberg, J. E., & Dominik, J., (2014). Extremely elevated methyl mercury levels in water, sediment and organisms in a Romanian reservoir affected by release of mercury from a chlor-alkali plant. *Water Research, 49*(1), 391–405.
9. Bravo, A. G., Loizeau, J. L., Dranguet, P., Makri, S., Björn, E., Ungureanu, V. G., & Cosio, C., (2016). Persistent Hg contamination and occurrence of Hg-methylating transcript (hgcA) downstream of a chlor-alkali plant in the Olt River (Romania). *Environmental Science and Pollution Research, 23*(1), 10529–10541.
10. Brinkmann, T., Santonja, G. G., Schorcht, F., Roudier, S., & Sancho, L. D., (2014). Best available techniques (BAT) reference document for the production of chlor-alkali. *JRC (Joint Research Center EU), JRC91156, 1*, pp. 5–10.

11. Burney, H. S., (1993). Membrane chlor-alkali process. *Modern Aspects of Electrochemistry, 24*(1), 393–438.
12. Busto, Y., Cabrera, X., Tack, F. M. G., & Verloo, M. G., (2011). Potential of thermal treatment for decontamination of mercury containing wastes from chlor-alkali industry. *Journal of Hazardous Materials, 186*(1), 114–118.
13. Cachada, A., Rodrigues, S. M., Mieiro, C., Ferreira da Silva, E., Pereira, E., & Duarte, A. C., (2009). Controlling factors and environmental implications of mercury contamination in urban and agricultural soils under a long-term influence of a chlor-alkali plant in the North–West Portugal. *Environmental Geology, 57*(1), 91–98.
14. Campanari, S., Guandalini, G., Coolegem, J., Ten Have, J., Hayes, P., & Pichel, A. H., (2019). Modeling, development, and testing of a 2 MW polymeric electrolyte membrane fuel cell plant fueled with hydrogen from a chlor-alkali industry. *Journal of Electrochemical Energy Conversion and Storage, 16*(4), 041001.
15. Collings, A. F., Hall, D. C., McCool, M. A., & Woolf, L. A., (1971). A diaphragm cell for diffusion measurements in liquids under pressure. *Journal of Physics E: Scientific Instruments, 4*(12), 1019.
16. Crook, J., & Mousavi, A., (2016). The chlor-alkali process: A review of history and pollution. *Environmental Forensics, 17*(3), 211–217.
17. Cussler Jr, E. L., & Dunlop, P. J., (1966). An experimental comparison of the Gouy and the diaphragm cell methods for studying isothermal ternary diffusion. *The Journal of Physical Chemistry, 70*(6), 1880–1888.
18. Dias, A. C. D. B. V., (2010). *Chlor-Alkali Membrane Cell Process: Study and Characterization* (Vol. 1, pp. 2–6). Doctoral dissertation, Universidade do Porto (Portugal)).
19. Dibble, G. J., & White, I. F., (1990). Practical experience in mercury and diaphragm cell conversions to membrane technology. *Modern Chlor-Alkali Technology, 4*(1), 291–305.
20. Du, R., Luo, X., Huang, Y., Chen, L., Huang, Z., Mao, X., & Wang, P., (2022). Polychlorinated dibenzo-p-dioxins and dibenzofurans in lotus from a lake historically polluted by the chlor-alkali industry: Occurrence, organ distribution and health risk from dietary intake. *Environmental Pollution, 292*(1), 118395.
21. Dufault, R., LeBlanc, B., Schnoll, R., Cornett, C., Schweitzer, L., Wallinga, D., & Lukiw, W. J., (2009). Mercury from chlor-alkali plants: Measured concentrations in food product sugar. *Environmental Health, 8*(1), 1–6.
22. Dullien, F. A. L., (1958). *Reconsideration of the Diaphragm Cell Method of*

Measuring Diffusion Coefficients (Vol. 1, pp. 3–9). Doctoral dissertation, University of British Columbia.

23. Esbri, J. M., Baselga, L., & Higueras, P., (2009). Evaluation of mercury dispersion from chlor-alkali industries in Spain. Environmental engineering and management. *Atiner, Atenas, 1*, 121–128.
24. Fell, C. J., & Hutchison, H. P., (1971). Use of diaphragm cell at elevated temperatures. *Industrial and Engineering Chemistry Fundamentals, 10*(2), 303–307.
25. Florkiewicz, T. F., (2008). Advanced diaphragm cell technology. *Modern Chlor-Alkali Technology, 8*(1), 165.
26. Foller, P. C., & Gu, J., (2016). Monitoring of diaphragm cell chlor-alkali plants for improved safety and performance. In: *Electrochemical Society Meeting Abstracts 230* (Vol. 1, No. 24, pp. 2–6, 1755–1755). The Electrochemical Society, Inc.
27. Gai, X. L., Li, X. H., Zhang, B., Xing, T., Zhao, J. J., & Ma, Z. H., (2016). Experimental study on sound absorption performance of microperforated panel with membrane cell. *Applied Acoustics, 110*(1), 241–247.
28. Garcia-Herrero, I., Margallo, M., Onandía, R., Aldaco, R., & Irabien, A., (2017). Life cycle assessment model for the chlor-alkali process: A comprehensive review of resources and available technologies. *Sustainable Production and Consumption, 12*(1), 44–58.
29. Garcia-Herrero, I., Margallo, M., Onandía, R., Aldaco, R., & Irabien, A., (2017). Environmental challenges of the chlor-alkali production: Seeking answers from a life cycle approach. *Science of the Total Environment, 580*(1), 147–157.
30. Guandalini, G., Foresti, S., Campanari, S., Coolegem, J., & Ten Have, J., (2017). Simulation of a 2 MW PEM fuel cell plant for hydrogen recovery from chlor-alkali industry. *Energy Procedia, 105*(1), 1839–1846.
31. Guney, M., Akimzhanova, Z., Kumisbek, A., Kismelyeva, S., Guney, A., Karaca, F., & Inglezakis, V., (2021). Assessment of distribution of potentially toxic elements in different environmental media impacted by a former chlor-alkali plant. *Sustainability, 13*(24), 13829.
32. Han, F., Li, W., Yu, F., & Cui, Z., (2014). Industrial metabolism of chlorine: A case study of a chlor-alkali industrial chain. *Environmental Science and Pollution Research, 21*(1), 5810–5817.
33. Harris, M., (2008). Phase-out issues for mercury cell technology in the chlor-alkali industry. *Modern Chlor-Alkali Technology, 8*(1), 19.
34. Heaton, C. A., (1993). The chemical industry. *Springer Science and*

Business Media, 2(1), 4–8.

35. Hławiczka, S., & Fudała, J., (2008). Assessment of atmospheric mercury emission reduction measures relevant for application in Poland. *Environmental Engineering Science, 25*(2), 163–172.

36. Hofmann, M., Müller, R., Christidis, A., Fischer, P., Klaucke, F., Vomberg, S., & Tsatsaronis, G., (2022). Flexible and economical operation of chlor-alkali process with subsequent polyvinyl chloride production. *AIChE Journal, 68*(1), e17480.

37. Horowitz, H. M., Jacob, D. J., Amos, H. M., Streets, D. G., & Sunderland, E. M., (2014). Historical mercury releases from commercial products: Global environmental implications. *Environmental Science and Technology, 48*(17), 10242–10250.

38. Hou, M., Chen, L., Guo, Z., Dong, X., Wang, Y., & Xia, Y., (2018). A clean and membrane-free chlor-alkali process with decoupled Cl_2 and H_2/NaOH production. *Nature Communications, 9*(1), 438.

39. Hu, J. S., Dong, J. L., Wang, Y., & Ren, X. D., (2012). Dye wastewater treatment by means of electrochemistry in diaphragm electrolyzer. *In Applied Mechanics and Materials* (Vol. 178, pp. 557–561). Trans Tech Publications Ltd.

40. Huber, T. B., & Benzing, T., (2005). The slit diaphragm: a signaling platform to regulate podocyte function. *Current Opinion in Nephrology and Hypertension, 14*(3), 211–216.

41. Hutchison, D. J., (2008). The chlor-alkali business. *Modern Chlor-Alkali Technology, 8*(1), 1–5.

42. Jacob, M. C., Favre, M., & Bensa, J. C., (1991). Membrane cell permeabilization with saponin and multiparametric analysis by flow cytometry. *Cytometry: The Journal of the International Society for Analytical Cytology, 12*(6), 550–558.

43. Karlsson, R. K., & Cornell, A., (2016). Selectivity between oxygen and chlorine evolution in the chlor-alkali and chlorate processes. *Chemical Reviews, 116*(5), 2982–3028.

44. Khasawneh, H., Saidan, M. N., & Al-Addous, M., (2019). Utilization of hydrogen as clean energy resource in chlor-alkali process. *Energy Exploration and Exploitation, 37*(3), 1053–1072.

45. Kim, M. I., Kim, N. J., Shang, L., Chang, Y., Lee, S., & Chang, H. N., (2009). Continuous production of succinic acid using an external membrane cell recycle system. *Journal of Microbiology and Biotechnology, 19*(11), 1369–1373.

46. Kinsey, J. S., Anscombe, F. R., Lindberg, S. E., & Southworth, G. R., (2004). Characterization of the fugitive mercury emissions at a chlor-alkali plant: overall study design. *Atmospheric Environment, 38*(4), 633–641.
47. Knepper, C., Savory, E. A., & Day, B., (2011). Arabidopsis NDR1 is an integrin-like protein with a role in fluid loss and plasma membrane-cell wall adhesion. *Plant Physiology, 156*(1), 286–300.
48. Laity, R. W., & Miller, M. P., (1964). Diaphragm-cell method for constant mass diffusion measurements. interionic diffusion and friction coefficients in molten silver nitrate-sodium nitrate mixtures. *The Journal of Physical Chemistry, 68*(8), 2145–2150.
49. Lakshmanan, S., & Murugesan, T., (2014). The chlor-alkali process: Work in progress. *Clean Technologies and Environmental Policy, 16*(1), 225–234.
50. Leddy, J. J., (1980). The chlor-alkali industry. *Journal of Chemical Education, 57*(9), 640.
51. Lee, D. Y., Elgowainy, A. A., & Dai, Q., (2017). *Life Cycle Greenhouse Gas Emissions of By-Product Hydrogen from Chlor-Alkali Plants (No. ANL/ESD-17/27)* (Vol. 4, No. 1, pp. 5–9). Argonne National Lab (ANL), Argonne, IL (United States).
52. Lee, D. Y., Elgowainy, A., & Dai, Q., (2018). Life cycle greenhouse gas emissions of hydrogen fuel production from chlor-alkali processes in the United States. *Applied Energy, 217*(1), 467–479.
53. Lei, T., Zhang, X., Jung, J., Cai, Y., Hou, X., Zhang, Q., & Qiao, J., (2018). Continuous electroreduction of carbon dioxide to formate on Tin nanoelectrode using alkaline membrane cell configuration in aqueous medium. *Catalysis Today, 318*(1), 32–38.
54. Li, C., Zhu, L., & Fleiter, T., (2014). Energy efficiency potentials in the chlor-alkali sector: A case study of Shandong province in China. *Energy and Environment, 25*(3–4), 661–686.
55. Li, K., & Li, N., (1993). Removal of carbon dioxide from breathing gas mixtures using an electrochemical membrane cell. *Separation Science and Technology, 28*(4), 1085–1090.
56. Li, K., Fan, Q., Chuai, H., Liu, H., Zhang, S., & Ma, X., (2021). Revisiting chlor-alkali electrolyzers: From materials to devices. *Transactions of Tianjin University, 27*(1), 202–216.
57. Margallo, M., Onandía, R., Aldaco, R., & Irabien, A., (2016). When life cycle thinking is necessary for decision making: Emerging cleaner technologies in the chlor-alkali industry. *Chemical Engineering Transactions, 52*(1), 475–480.

58. Masding, P. W., & Browning, N. D., (2008). A dynamic model of a mercury chlorine cell. *Modern Chlor-Alkali Technology, 8*(1), 247–259.
59. McCluney, S. A., & Van Zee, J. W., (1989). An optimization analysis of a diaphragm cell/evaporator system for NaOH production. *Journal of the Electrochemical Society, 136*(9), 2556.
60. Melián-Martel, N., Sadhwani, J. J., & Báez, S. O. P., (2011). Saline waste disposal reuse for desalination plants for the chlor-alkali industry: The particular case of Pozo Izquierdo SWRO desalination plant. *Desalination, 281*(1), 35–41.
61. Mihaiescu, T., Mihaiescu, R., & Odagiu, A., (2012). Environmental issues within the chlor-alkali manufacturing industry-mercury cell process. *Bull. Univ. Agric. Sci. Vet. Med., 69*(1), 168–176.
62. Mo, H., & Ng, H. Y., (2010). An experimental study on the effect of spacer on concentration polarization in a long channel reverse osmosis membrane cell. *Water Science and Technology, 61*(8), 2035–2041.
63. Mo, H., & Ng, H. Y., (2010). Organic fouling development in a long channel RO membrane cell. *Water Science and Technology: Water Supply, 10*(4), 672–678.
64. Moorhouse, J., (2008). *Modern Chlor-Alkali Technology* (Vol. 1, 8, pp. 5–10). John Wiley & Sons.
65. Morris, D. R., (1991). Exergy analysis and cumulative exergy consumption of complex chemical processes: the industrial chlor-alkali processes. *Chemical Engineering Science, 46*(2), 459–465.
66. Morris, G. J., (1990). Improving membrane cell operation. *Modern Chlor-Alkali Technology, 4*, 197–213.
67. Moussallem, I., Jörissen, J., Kunz, U., Pinnow, S., & Turek, T., (2008). Chlor-alkali electrolysis with oxygen depolarized cathodes: History, present status, and future prospects. *Journal of Applied Electrochemistry, 38*(1), 1177–1194.
68. Muñoz, E., & Navia, R., (2011). Life cycle assessment of solid waste management strategies in a chlor-alkali production facility. *Waste Management and Research, 29*(6), 634–643.
69. Nidola, A., (1986). Zero gap membrane cell and SPE cell technologies vs. current density scale up. *Membranes and Membrane Processes, 3*(2), 281–298.
70. O'Brien, T. F., Bommaraju, T. V., Hine, F., O'Brien, T. F., Bommaraju, T. V., & Hine, F., (2005). Overview of the chlor-alkali industry. *Handbook of Chlor-Alkali Technology: Volume I: Fundamentals, Volume II: Brine*

Treatment and Cell Operation, Volume III: Facility Design and Product Handling, Volume IV: Plant Commissioning and Support Systems, Volume V: Corrosion, Environmental Issues, and Future Development, 1, 37–74.

71. Olufemi, B. A., Kehinde, A. J., & Ogboja, O., (2016). Operational and performance optimization of an expanded area diaphragm cell. *Journal of Engineering Research, 21*(1), 41–52.
72. Pirrone, N., Cinnirella, S., Feng, X., Finkelman, R. B., Friedli, H. R., Leaner, J., & Telmer, K., (2010). Global mercury emissions to the atmosphere from anthropogenic and natural sources. *Atmospheric Chemistry and Physics, 10*(13), 5951–5964.
73. Prout, N. M., & Moorhouse, J. S., (2012). *Modern Chlor-Alkali Technology* (Vol. 4, pp. 6–9). Springer Science & Business Media.
74. Reig, M., Casas, S., Aladjem, C., Valderrama, C., Gibert, O., Valero, F., & Cortina, J. L., (2014). Concentration of NaCl from seawater reverse osmosis brines for the chlor-alkali industry by electrodialysis. *Desalination, 342*(1), 107–117.
75. Reis, A. T., Rodrigues, S. M., Araújo, C., Coelho, J. P., Pereira, E., & Duarte, A. C., (2009). Mercury contamination in the vicinity of a chlor-alkali plant and potential risks to local population. *Science of the Total Environment, 407*(8), 2689–2700.
76. Robinson Jr, R. L., Edmister, W. C., & Dullien, F. A. L., (1965). Calculation of diffusion coefficients from diaphragm cell diffusion data. *The Journal of Physical Chemistry, 69*(1), 258–261.
77. Rodrigues, M., De Mattos, T. T., Sleutels, T., Ter Heijne, A., Hamelers, H. V., Buisman, C. J., & Kuntke, P., (2020). Minimal bipolar membrane cell configuration for scaling up ammonium recovery. *ACS Sustainable Chemistry and Engineering, 8*(47), 17359–17367.
78. Rosales-Huamani, J. A., Medina-Collana, J. T., Diaz-Cordova, Z. M., & Montaño-Pisfil, J. A., (2021). Factors influencing the formation of sodium hydroxide by an ion exchange membrane cell. *Batteries, 7*(2), 34.
79. Roy, H., Barua, S., Ahmed, T., Mehnaz, F., Islam, M. S., & Mujtaba, I. M., (2022). A sustainable integration approach of chlor-alkali industries for the production of PVC and clean fuel hydrogen: Prospects and Bangladesh perspectives. *Processes, 10*(8), 1638.
80. Satoh, D., Hirose, T., Harita, Y., Daimon, C., Harada, T., Kurihara, H., & Ohno, S., (2014). aPKCλ maintains the integrity of the glomerular slit diaphragm through trafficking of nephrin to the cell surface. *The Journal of Biochemistry, 156*(2), 115–128.

81. Schoenberger, H., (1990). The world chlor-alkali industry past–present–future. *Modern Chlor-Alkali Technology, 4*, 1–15.
82. Selvaraj, H., Chandrasekaran, K., & Gopalkrishnan, R., (2016). Recovery of solid sulfur from hydrogen sulfide gas by an electrochemical membrane cell. *RSC Advances, 6*(5), 3735–3741.
83. Snyder, L. D., Miller, N. H., & Stavins, R. N., (2003). The effects of environmental regulation on technology diffusion: The case of chlorine manufacturing. *American Economic Review, 93*(2), 431–435.
84. Song, Z., Li, P., Ding, L., Li, Z., Zhu, W., He, T., & Feng, X., (2018). Environmental mercury pollution by an abandoned chlor-alkali plant in Southwest China. *Journal of Geochemical Exploration, 194*(1), 81–87.
85. Soto, D. X., Roig, R., Gacia, E., & Catalan, J., (2011). Differential accumulation of mercury and other trace metals in the food web components of a reservoir impacted by a chlor-alkali plant (Flix, Ebro River, Spain): Implications for biomonitoring. *Environmental Pollution, 159*(6), 1481–1489.
86. Stanley, K., (2008). Practical operating differences in converting a diaphragm cell chlor-alkali plant to a membrane electrolyzer plant. *Modern Chlor-Alkali Technology, 8*, 182.
87. Suleman, F., Dincer, I., & Agelin-Chaab, M., (2016). Comparative impact assessment study of various hydrogen production methods in terms of emissions. *International Journal of Hydrogen Energy, 41*(19), 8364–8375.
88. Sznopek, J. L., & Goonan, T. G., (2000). *The Materials Flow of Mercury in the Economies of the United States and the World* (Vol. 1, pp. 2–6.). US Department of the Interior, US Geological Survey
89. Thiel, G. P., Kumar, A., Gómez-González, A., & Lienhard, J. H., (2017). Utilization of desalination brine for sodium hydroxide production: technologies, engineering principles, recovery limits, and future directions. *ACS Sustainable Chemistry and Engineering, 5*(12), 11147–11162.
90. Toma, K., Seshima, F., Maruyama, A., Arakawa, T., Yano, K., & Mitsubayashi, K., (2019). Improved performance Air bio-battery based on efficient oxygen supply with a gas/liquid highly porous diaphragm cell. *Biosensors and Bioelectronics, 124*(1), 253–259.
91. Tonini, D. R., Gauvin, D. A., Soffel, R. W., & Freeman, W. P., (2003). Achieving low mercury concentrations in chlor-alkali wastewaters. *Environmental Progress, 22*(3), 167–173.
92. Ullrich, S. M., Ilyushchenko, M. A., Tanton, T. W., & Uskov, G. A., (2007). Mercury contamination in the vicinity of a derelict chlor-alkali plant: Part II: Contamination of the aquatic and terrestrial food chain and

potential risks to the local population. *Science of the Total Environment, 381*(1–3), 290–306.

93. Ursua, A., Gandia, L. M., & Sanchis, P., (2011). Hydrogen production from water electrolysis: Current status and future trends. *Proceedings of the IEEE, 100*(2), 410–426.
94. Verma, S., Suresh, S., Mili, M., Diwedi, P., Bajpai, H., Khan, M. A., & Srivastava, A. K., (2022). Future process options for brine sludge management from the chlor-alkali industry. *International Journal of Environmental Science and Technology*, 1–20.
95. von Weymarn, N., Kiviharju, K., & Leisola, M., (2002). High-level production of D-mannitol with membrane cell-recycle bioreactor. *Journal of Industrial Microbiology and Biotechnology, 29*, 44–49.
96. Walker, T. R., (2016). Mercury concentrations in marine sediments near a former mercury cell chlor-alkali plant in eastern Canada. *Marine Pollution Bulletin, 107*(1), 398–401.
97. Wang, L., Tan, Y., Gao, L., Lei, J., Chen, C., & Shi, Y., (2020). Effect of astragaloside on diaphragm cell apoptosis in chronic obstructive pulmonary disease. *Food Science and Nutrition, 8*(12), 6357–6366.
98. Wang, X., Teichgraeber, H., Palazoglu, A., & El-Farra, N. H., (2014). An economic receding horizon optimization approach for energy management in the chlor-alkali process with hybrid renewable energy generation. *Journal of Process Control, 24*(8), 1318–1327.
99. Wang, X., Tong, C., Palazoglu, A., & El-Farra, N. H., (2014). Energy management for the chlor-alkali process with hybrid renewable energy generation using receding horizon optimization. *In: 53rd IEEE Conference on Decision and Control* (pp. 4838–4843). IEEE.
100. Xia, Y., Ouyang, M., Yufit, V., Tan, R., Regoutz, A., Wang, A., & Brandon, N. P., (2022). A cost-effective alkaline polysulfide-air redox flow battery enabled by a dual-membrane cell architecture. *Nature Communications, 13*(1), 2388.
101. Yamaguchi, K., (1990). Modification of diaphragm-type chlor-alkali cells to ion exchange membrane cells. *Journal of Applied Electrochemistry, 20*, 656–661.
102. Zhang, L. N., Li, R., Zang, H. Y., Tan, H. Q., Kang, Z. H., Wang, Y. H., & Li, Y. G., (2021). Advanced hydrogen evolution electrocatalysts promising sustainable hydrogen and chlor-alkali co-production. *Energy and Environmental Science, 14*(12), 6191–6210.

CHAPTER

2 Significance of Chlorine and Caustic

Contents

2.1. INTRODUCTION

Chlorine & sodium hydroxide, also known as caustic soda, are prominent chemicals in global production and play a significant role in manufacturing various everyday products. The items mentioned above comprise a range of commodities such as detergents, deodorants, pharmaceuticals, herbicides, disinfectants, pesticides, & plastics (Caldwell, 1981).

In 1774, Carl Wilhelm Scheele made an initial observation regarding the potential use of chlorine. Specifically, Scheele noted that chlorine had a bleaching effect on vegetable matter. This discovery was made during Scheele's investigation into reactivity of greenish-yellow gas produced when hydrochloric acid & pyrolusite (manganese dioxide) react. Production of toxic and asphyxiating greenish-yellow chlorine vapors has been documented since 13th century, coinciding with chemists' discovery of aqua regia. Around 1668 a flatus incoercible, characterized by a forcible chlorine release, was produced due to the reaction between ammonium chloride and nitric acid. In 1785, Berthollet attempted to substitute the conventional solar bleaching method, also known as crofting, OR grassing, for textiles with elemental chlorine but was unsuccessful (Martin, 1961). Presence of elemental chlorine resulted in worker discomfort, corrosion of metallic components, and degradation of textile materials. The historical employment of chlorine in the formation of potassium hypochlorite, which involves the absorption of chlorine in a potash solution for the purpose of large-scale industrial bleaching, can be dated back to the year 1789. This pioneering development was attributed to McPherson. In 1798, Renant obtained a patent for producing calcium hypochlorite solution by substituting milk of lime for the potash solution. Tennant was granted the initial patent for bleaching powder in 1789. In 1808, Sir Humphry Davy identified chlorine as an element and bestowed upon it the name "chlorine," which is derived from the Greek word for green. Chlorine is a greenish-yellow gas (Sharma et al., 1997).

Did You Know?

Chlorine gas has a distinct, pungent odor that is often associated with swimming pools.

The pivotal advancements in the field of bleaching involved the utilization of chlorine for chemical bleaching and the identification of calcium hypochlorite bleaching powder as a viable means of transporting chlorine. The Industrial Revolution, characterized by population growth and heightened demand for textiles, witnessed a significant impact of various technologies on textile bleaching operations in Europe & Great Britain. Advent of the power loom facilitated production of textiles. Nonetheless, the process of bleaching through laying out the fabric in exposed fields necessitated a prolonged duration of time. It incurred escalating costs due to the surging value of land. The implementation of the chlorine bleaching process resulted in a reduction of operational time from several months to a mere few days. Additionally, this process facilitated the liberation of extensive regions for more efficient utilization. The pulp and paper industry adopted the use of bleaching powder due to its significant enhancement in textile bleaching efficiency (King & Hancock, 1996).

In the wake of the revelation that bacteria were the causative agents of certain diseases, several researchers conducted investigations into sewage and drinking water chlorination during the 1890s. By the year 1912, using chlorine for water treatment became customary practice, resulting in a noteworthy decrease in the prevalence of waterborne illnesses such as typhoid. "Winter Typhoid" incidence in Montreal (Canada) was documented to be 549 cases between October and December 1909. However, following the implementation of drinking water chlorination in 1910, the number of reported cases during the same period decreased significantly to 170. Between 1756 and 1932, the pulp-making industry experienced a significant advancement with the adoption of chlorine (Riffat & Cheng, 2009). The utilization of hypochlorite, a variant of chlorine, was found to be efficacious in the removal of color or color-producing substances from cellulosic fibers, while simultaneously mitigating the degradation of the fibers. During 19th century, most chlorine produced was utilized by water disinfection operations and paper and pulp industry. Chlorine was initially employed for disinfection in medical facilities in 1823, marking its inaugural application. In 1826, water of chlorine was utilized in obstetric wards as a preventative measure against puerperal fever. Additionally, during widespread European cholera epidemic, fumigation with chlorine was a common practice (Ghannadzadeh & Sadeqzadeh, 2017).

Dichloroethane was first synthesized in 1795, while chloroform was produced through chemical synthesis in 1831. The anesthetic properties of chloroform had been acknowledged and integrated into surgical procedures by 1848.

The utilization of chlorine for purpose of water purification amidst the typhoid outbreak in Niagara Falls in 1912 represented a noteworthy achievement for the field. It is noteworthy that in 1897, bleaching powder was utilized to disinfect contaminated water sources during a typhoid outbreak in England. During World War I, United States' chemical industry gain independence from its European counterpart, leading to the development of numerous novel applications of chlorine (Nikravesh et al., 1995).

Several novel applications of chlorine, such as the production of chlorinated solvents, ethylene glycol, vinyl chloride, & various others, were formulated during the period spanning from 1920 to 1940. The utilization of chlorine for military purposes experienced a surge during the war era, leading to the emergence of novel applications. This trend persisted and resulted in additional commodities for non-military use in the aftermath of World War II. The advancements made in synthetic organic chemistry during 19th century facilitated the creation of alternative options for natural products & novel and practical compounds, encompassing both intermediate and end products. Due to its chemical reactivity, distinctive characteristics, and cost-effectiveness, chlorine was employed in various applications, such as pharmaceuticals, solvents, & dyes (Sharma et al., 1996).

Chlorine and caustic substances are commonly employed in producing various goods widely utilized across diverse domains. The subsequent segments delineate certain chlorine, potassium hydroxide (KOH) and caustic soda applications. The interested party is directed to consult references for a comprehensive inventory of the resultant products stemming from the interaction of chlorine and caustic. These sources contain a compilation of the intermediate compounds generated from chlorine, the ultimate products, and their respective applications (Arevalo et al., 2006).

2.2. END USES OF CHLORINE

Chlorine finds application in producing organic and inorganic chemicals through direct utilization or indirect involvement.

Chlorine finds direct application in paper and pulp manufacturing and in operations related to water treatment. Chlorine is utilized within paper and pulp sector for pulp bleaching, which results in the production of a superior whitened material that is free from dark lignin & other unwanted residuals. Municipalities and other organizations continue to use chlorine as a disinfectant to treat wastewater, process water, and potable water streams. Using chlorine has led to eliminating waterborne illnesses like typhoid and cholera in developed nations. Chlorine can eliminate hydrogen sulfide, iron compounds, & organic substances which contribute to the unpleasant taste or odor commonly associated with water (Evan, 2005).

Did You Know?

Chlorine was first used as a disinfectant in the 19th century by John Snow, to combat the spread of cholera in London.

2.2.1. Organic Chemicals

Chlorine finds its primary application in producing organic compounds, with the most extensively manufactured product being polyvinyl chloride (PVC). The production of PVC entails the application of vinyl chloride monomer (VCM) and ethylene dichloride (EDC), as illustrated by Eqns. (1) and (2), correspondingly. PVC is highly adaptable thermoplastic material. In its rigid form, PVC finds application in various industries such as construction, packaging, and consumer goods. Its uses include but are not limited to siding, windows, piping, bottles, & packaging sheeting. Plasticized PVC exhibits flexibility and is utilized in various applications such as floor coverings, sheet & film production, cable & wire insulation, & numerous other industrial uses (Harrison et al., 2006). PVC is frequently employed judiciously as a replacement for metallic materials.

$$CH_2 = CH_2 + Cl_2 \rightarrow CH_2CI - CH_2CI \quad (1)$$

$$CH_2Cl - CH_2Cl \rightarrow CH_2 = CHCI + HCL \quad (2)$$

A chemical molecule called propylene oxide is created when propylene reacts with chlorine to create propylene chlorohydrin. This intermediate product is then subjected to dehydrochlorination using either caustic soda or lime, as shown in Eqns. (3)–(5). The resulting propylene oxide is utilized in manufacturing polyether polyols, which are employed in the production of urethane foam. Additionally, it is utilized in producing unsaturated polyester resins when combined with propylene glycol & in food & pharmaceutical sectors. A chemical molecule called epichlorohydrin (EPI) is created by chlorinating propylene to create allyl chloride, then dehydrochlorinating it (as described in Eqns. (6) and (7)). EPI finds its application in the production of epoxy resins, which are utilized in the manufacturing of fiber-reinforced composites, laminates, protective coatings, & adhesives (Calvet, 1989).

$$Cl_2 + H_2O \longrightarrow HOCl + HCl \quad (3)$$

$$\underset{\text{Propylene}}{CH_3{-}CH{=}CH_2} + HOCl \longrightarrow \underset{\text{Propylene chlorohydrin}}{CH_3{-}\underset{OH}{CH}{-}\underset{Cl}{CH_2}} \quad (4)$$

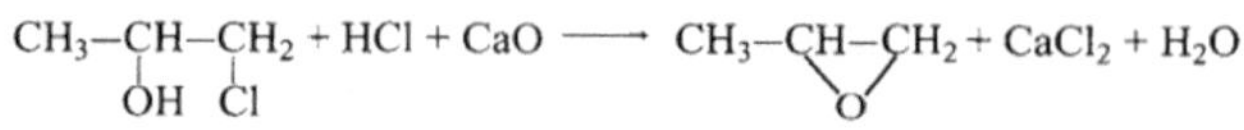

Propylene Chlorohydrin   Propylene Oxide   (5)

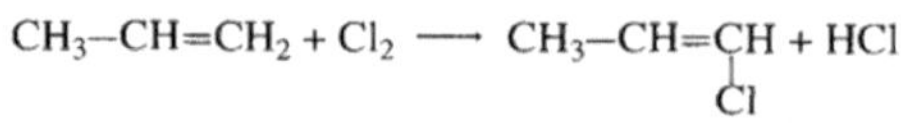

Propylene   Allyl Chloride   (6)

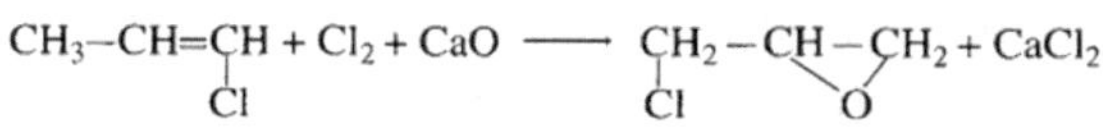

Allyl Chloride   Epichlorohydrin   (7)

The carbonylation agent, phosgene, is generated through the chlorination of carbon monoxide. This agent converts amines to isocyanates, as evidenced by Eqns. (8) and (9). Chemical reaction between phosgene and diphenylmethane diamine yields methylene diphenyl diisocyanate (MDI), while the reaction with toluene diamine produces toluene diisocyanate (TDI). Isocyanates are utilized in manufacturing polyurethanes, which find applications in the construction and automotive sectors for producing rigid & flexible foams, coatings, elastomers, & adhesives (Wania & Mackay, 1995).

$$CO + Cl_2 \rightarrow COCl_2 \qquad (8)$$

$$R\text{–}NH_2 + COCl_2 \rightarrow R\text{–}N{=}C{=}O + 2HCl \qquad (9)$$

Amine   Phosgene   Isocyanate

1,1,1-trichloroethane, perchloroethylene (PCE), and trichloroethylene (TCE) are chlorinated solvents from EDC and VCM. TCE and 1,1,1-trichloroethane are extensively used in vapor degreasing procedures, while PCE is frequently used in dry cleaning. Chlorofluorocarbons (CFCs) are synthesized through the reaction of HF with carbon tetrachloride (CCl_4) to produce CFC-11 and -12 and with PCE to yield CFC-113, -114, and -115. These compounds are commonly used as blowing agents, solvents, & refrigerants in household & commercial air conditioning systems. HCFC-22, a hydrochlorofluorocarbon, is a compound that is less detrimental to ozone layer than CFCs. It's synthesized from chloroform & serves not only as a refrigerant in air-conditioning systems but also as an intermediary in production of tetrafluoroethylene, which is utilized in the manufacturing of polytetrafluoroethylene and other fluoropolymers. HCFC-142b, a chemical compound obtained from 1,1,1-trichloroethane, functions as a blowing agent for polystyrene and a key component for vinylidene fluoride, which is employed in the production of fluoroplastics and fluoroelastomers (Mackay & Peterson, 1991).

HFC-134a is viable alternative to CFC-12 for manufacturing of refrigerants due to its lack of chlorine and environmentally friendly properties. This substance is derived from intermediates of organic compounds that have undergone chlorination. The chemical reactions involved in the synthesis of fluorinated compounds from CCl_4, chloroform ($CHCl_3$), and tetrachloroethylene ($CCl_2{=}CCl_2$) are depicted in Eqns. (10) through (18). Figure 2.1 illustrates the fluorinated compounds that have been synthesized utilizing C-2 compounds (Alexander, 1985).

Methyl chloride finds its application in manufacturing chlorosilanes, which serve as crucial intermediates in producing silicone-based fluids, elastomers, and resins. Methylene chloride is utilized in various industries, such as film processing, paint removal, metal cleaning, urethane foam expansion, & electronic & pharmaceutical sectors. Household food wraps, including Saran, comprise VCM and vinylidene chloride. The latter compound is synthesized through the chlorination of EDC to yield 1, 1, 2-trichloroethane, which is subsequently subjected to dehydrochlorination using caustic soda (Plato & Glasgow, 1969).

CCl_4

$$+HF \rightarrow CCl_3F + HCl \quad (10)$$
(Trichlorofluoromethane: CFC-11)

$$+2HF \rightarrow CCl_2F_2 + 2HCl \quad (11)$$
(Dichlorodifluoromethane: CFC-12)

$$+3HF \rightarrow CClF_3 + 3HCl \quad (12)$$
(Chlorotrifluoromethane: CFC-13)

$CHCl_3$

$$+HF \rightarrow CHCl_2F + HCl \quad (13)$$
(Dichlorofluoromethane: HCFC-21)

$$+2HF \rightarrow CHClF_2 + 2HCl \quad (14)$$
(Chlorodifluoromethane: HCFC-22)

$$+3HF \rightarrow CHF_3 + 3HCl \quad (15)$$
(Trifluoromethane: HFC-23)

$CCl_2{=}CCl_2$

$$+Cl_2 + 3HF \rightarrow CCl_2F{-}CClF_2 + 3HCl \quad (16)$$
(1,1,2-Trichlorotrifluoroethane: CFC-113)

$$+Cl_2 + 4HF \rightarrow CClF_2{-}CClF_2 + 4HCl \quad (17)$$
(1,2-Dichlorotetrafluoroethane: CFC-114)

$$+Cl_2 + 5HF \rightarrow CF_3{-}CClF_2 + 5HCl \quad (18)$$
(Chloropentafluoroethane: CFC-115)

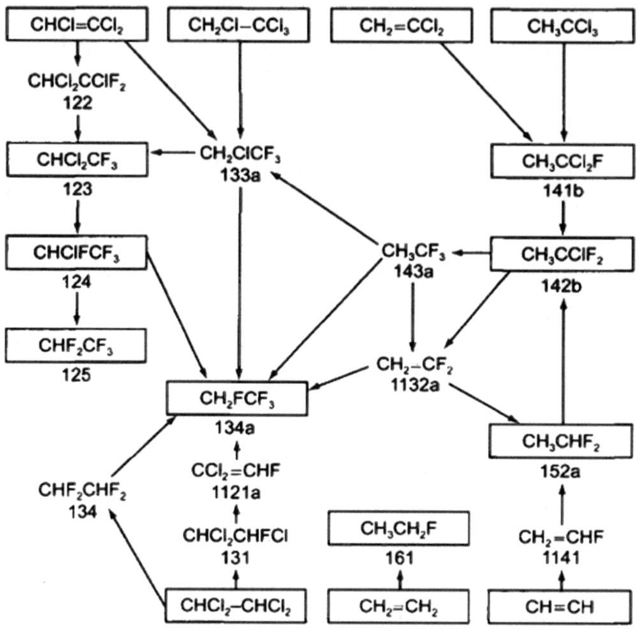

Figure 2.1. *Example of C_2-HCFCs, HFCS, and their forerunners.*

Source: Fumio Hine, Creative Commons License.

The process of chlorinating benzene can result in the production of chlorobenzenes when a metal catalyst, such as $FeCl_3$ or $AlCl_3$, is present. This reaction can be represented by Eqns. (19) and (20).

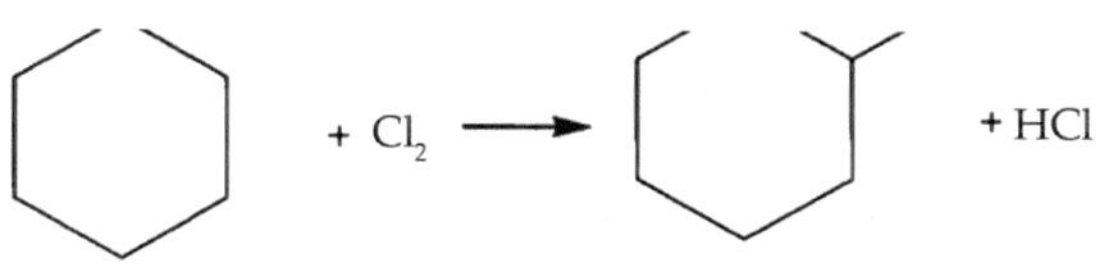

(19)

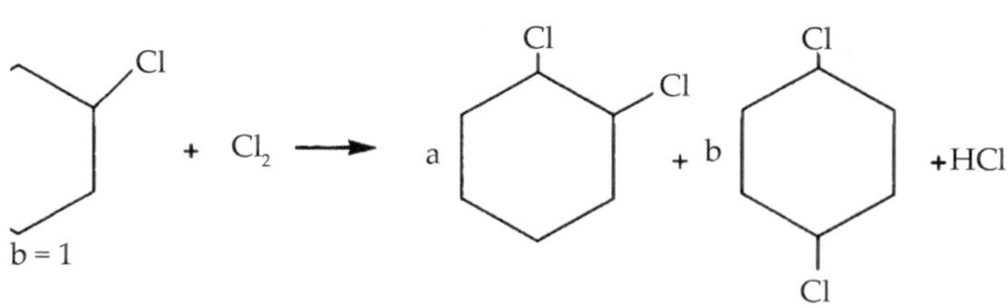

(20)

The utilization of monochlorobenzene is observed in nitrochlorobenzenes manufacturing, which serve as intermediates in producing various chemical compounds such as rubber chemicals, antioxidants, dyes, & pigments. Monochlorobenzene is also utilized to produce diphenyl ether, crucial constituent of heat transfer fluids. Dichlorobenzenes find application in producing mothballs, dye intermediates, & process solvents (Pan & Xing, 2008).

Raw material for production of allyl resins & explosives, such as nitroglycerine, is synthetic glycerin, compound in question is acquired via process of chlorination of allyl alcohol, resulting in formation of dichlorohydrin, which is subsequently subjected to hydrolysis. Furthermore, it is utilized as a humectant in the processing of tobacco and as a plasticizer in the manufacture of cellophane. Phosgene is employed as a carbonylating agent for bisphenol A (as shown in Eqn. (21)) to yield polycarbonate resins. These resins have diverse applications in domains including but not limited to automotive parts, appliances, power tools, and glazing (Gao et al., 1992).

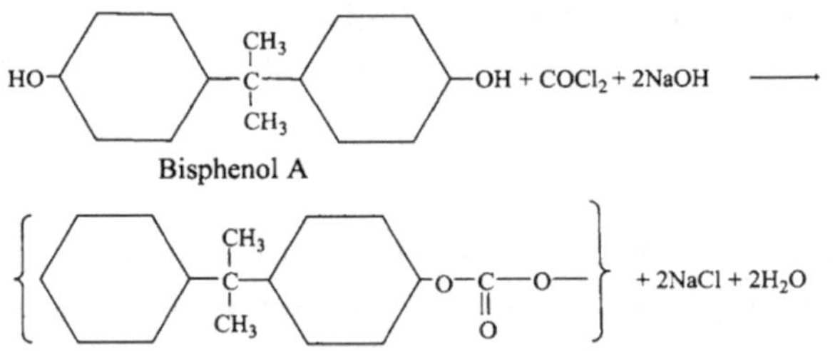

(21)

Polycarbonate

The production of chloroprene, a key component in neoprene manufacturing, involves reaction of chlorine with butadiene. Neoprene is an elastomer that exhibits exceptional resilience and resistance to various environmental factors, including ozone, flame, and weathering (Eqn. 22).

$$CH_2\text{= CH–CH=}\ CH_2\ Cl_2 \qquad CH_2\text{=CCl–CH=}\ CH_2 + \text{HCl} \quad (22)$$

Butadiene Chloroprene

Chlorine-based compounds such as methyl chloride, 1,1,1-Trichloroethane, and chlorobenzenes are utilized as intermediates in the manufacturing process of various pesticides, including herbicides, insecticides, fumigants, and fungicides. Pentachlorophenol serves as a wood preservative.

Within the C_{10}–C_{30} range, there exist numerous straight-chain saturated hydrocarbons that can be chlorinated to produce chlorinated paraffin. This substance is commonly utilized as a coolant, lubricant for metal-working compounds, plasticizer, and flame retardant (Huckins et al., 2006).

The chlorination process that is subject to regulation produces di- and tri-chloroisocyanuric acids with a chlorine concentration exceeding 90%.

The acids in question are highly esteemed for their robust disinfectant and bleaching properties and are frequently employed in industrial and commercial laundry facilities, as well as in dishwashing detergents. Dry packaging is a viable option for domestic use, and it presents a competitive alternative to the widely utilized sodium hypochlorite.

Cyanuric chloride is a versatile compound that finds application in producing herbicides, optical brighteners, pharmaceuticals, explosives, & surfactants. This compound is produced via chemical reaction between chlorine & hydrogen cyanide. Tetraethyl lead was produced using ethyl chloride, which is synthesized via the reaction between ethylene and HCl.

Numerous commodities entail the utilization of chlorine in their production procedures, such as polyurethanes, epoxy resins, refrigerants, process solvents, chlorinated solvents, cellophane, wood preservatives, flame retardants, herbicides, pharmaceuticals, & explosives (Gobas, 1993).

2.2.2. Inorganic Chemicals

Chlorine finds application in the manufacturing of numerous inorganic chemicals. Primary output of the process is titanium dioxide, which is synthesized through the reaction of rutile (92–96% TiO_2) or ilmenite blend (40–55% TiO_2) & rutile with chlorine to produce $TiCl_4$. This compound is subsequently refined and oxidized to yield TiO_2, a widely utilized pigment in producing paper, plastics, paints, & rubber.

Bromine is a chemical element that is frequently utilized in diverse industrial applications, including but not limited to fire extinguishing agents, flame retardants, drilling fluids (calcium and zinc bromides), and fumigants for soil and building pest control. The synthesis of bromine involves the reaction of hot brine containing bromide ion with chlorine. Hydrochloric acid of high purity is utilized in various industries such as food, electronics, and pharmaceuticals which is synthesized through the combustion of H_2 and $TiCl_2$. Moreover, diverse hydrocarbon chlorination methodologies retrieve a substantial quantity of gaseous hydrogen chloride or hydrochloric acid (Fawell, 1993).

The production of hypochlorites of calcium, lithium, sodium, & chlorinated trisodium phosphate entails the utilization of chlorine. Sodium hypochlorite, with a strength of approximately 5%, is widely utilized in the household bleach industry and for residential swimming pools. On the other hand, 15% sodium hypochlorite, classified as an industrial bleach, is employed for water treatment in municipal and industrial facilities. The chemical compound sodium hypochlorite is synthesized through a chemical reaction between the elements chlorine & caustic soda. Primary application of calcium hypochlorite is as chemical agent for swimming pools. Aluminum chloride is a catalyst that finds its application in producing alkyl & ethyl benzenes, ethyl chloride, dyestuffs, & hydrocarbon resin. It is synthesized from aluminum or alumina & chlorine. Additionally, it's utilized in manufacturing of pharmaceuticals & titanium dioxide (Godlewska, 2006).

The production of exotic metals, namely titanium and zirconium, involves the formation of their corresponding chlorides through a chemical reaction between chlorine and oxides. The production of metallic titanium involves the reaction of $TiCl_4$ with magnesium or molten sodium, whereas metallic zirconium is obtained through the reaction of ZrC_4 with magnesium. Magnesium metal and chlorine are produced through the process of molten salt electrolysis of magnesium chloride. The production of metallic sodium can be achieved through the process of electrolysis of a fused salt state mixture containing sodium chloride and calcium chloride (Elnabarawy et al., 1986).

2.3. END USES OF SODIUM HYDROXIDE

Range of applications for caustic soda is more diverse than that of chlorine. The widespread utilization of this substance can be attributed to its potent alkaline properties, which enable it to counteract acidic compounds effectively. Additionally, it exhibits reactivity towards certain metals and oxides, including aluminum or Al_2O_3, forming anionic aluminates.

Did You Know?

Caustic soda is used in the process of making soap through a reaction called saponification.

As elaborated below, caustic soda finds application in manufacturing of both organic and inorganic chemicals, either through direct or indirect means (Sellers & Prewitt, 1990).

Caustic finds direct application in various industries such as paper and pulp, soap & detergent production, petroleum & natural gas operations, & alumina manufacturing. Wood pulp is widely used in producing various products including paperboard and paper, rayon, and cellulose ester plastics. This material is derived from small wood chips sourced from hardwood or softwood logs. Producing wood pulp involves using white liquor & steam containing caustic soda & sodium sulfide to digest wood chips. After the breakdown of lignin & subsequent transformation of wood chips into pulp, resultant pulp is subjected to a washing process that separates it from the black liquor. This liquor is a blend of exhausted pulping chemicals, deteriorated lignin byproducts, and other substances. At this juncture, the pulp exhibits a dark brown hue and undergoes subsequent processing to eliminate residual lignin. The pulp is then subjected to a bleaching process utilizing chlorine or other oxidizing agents to achieve a brightened state, producing a superior grade of white paper (Hu et al., 2011).

Soaps are predominantly composed of sodium salts of fatty acids, accompanied by minor constituents such as fragrances, bleaching agents, and builders that form complexes with calcium and magnesium ions. Stearic acid is a common fatty acid found in fat, oleic acids, & coconut oil. Detergents, which represent a significant market for caustic, comprise various components such as surface-active agents, Sud regulators, corrosion inhibitors, & other additives. Surfactants are synthesized through the chemical reaction between caustic and linear alkyl sulfonates, alcohol sulfates, & ether sulfates. The chemical compound known as caustic soda is utilized in saponification, which involves the conversion of oils, fatty acids, and waxes into various consumer goods such as cosmetics and greases (Stolaroff et al., 2008).

The petroleum and natural gas industry utilizes caustic as an augmentative component to drilling mud, which is circulated to facilitate cooling and lubrication and mitigate the corrosive effects of acidic species. Including caustic has a dual effect of reducing the corrosion of drilling materials and improving petroleum recovery. The above-mentioned process involves the dissolution of drilling mud constituents under alkaline conditions, coupled with the mitigation of interfacial tension among oil, water, and rock. These mechanisms aid in the recovery of oil during chemical flooding operations. Furthermore, acidic properties serve as an agent that eradicates bacteria. Caustic is commonly employed in the oil and gas industry to eliminate H_2S from crude hydrocarbons and extract nonvolatile constituents, such as phenols, cresols, fatty acids, and naphthenic acids, from hydrocarbon streams. Cresols, which are employed in resins manufacturing, are obtained from the spent caustic solution that results from scrubbing petroleum distillates (Thomason et al., 2016).

The Bayer process employs caustic as a primary agent in alumina production from bauxite, making it a significant direct application of this chemical compound. The process of extracting alumina involves subjecting bauxite, which is an ore comprising hydrated alumina and impurities like iron and silica, to acidic treatment, which results in the solubilization of alumina in the form of sodium aluminate. The solution undergoes filtration to eliminate the presence of insoluble components. After introducing the sodium aluminate solution, hydrolysis, and cooling procedures are implemented to generate aluminum hydroxide. The resulting aluminum hydroxide is then subjected to calcination, ultimately yielding pure alumina. The production of metallic aluminum is achieved through the process of electrolysis (Phasey et al., 2017).

2.3.1. Organic Chemicals

Caustic soda is prevalent in producing organic chemicals, wherein it serves various purposes such as the neutralization of acids, pH regulation, off-gas scrubbing, and dehydrochlorination. Moreover, it can function as a catalyst and a sodium source during chemical reactions. The production of propylene oxide, a crucial constituent for the manufacturing of polyether polyols utilized in a diverse range of products such as flexible and rigid urethane foams, elastomers, coatings, and injection molding polymers, involves the dehydrochlorination process of propylene chlorohydrin utilizing lime or caustic soda. The utilization of caustic is observed in the production of EPI and the dehydrochlorination phase of epoxy resin manufacturing, which is employed to fabricate protective coatings, laminates, and fiber-reinforced plastics. The utilization of caustic in the hydrolysis of EPI leads to the production of glycerin, which finds application in various industries such as pharmaceuticals, tobacco, cosmetics, and food and beverage (Travis, 1990).

Chelating agents, which are employed to sequester metal ions such as calcium, magnesium, iron, and copper, are synthesized via a reaction between caustic and various organic acids, including ethylene diamine tetraacetic acid, hydroxy ethylenediamine triacetin acid, nitrilotriacetic acid, citric acid, gluconic acid, and diethylene triamine pentaacetic acid. Cellulose ether production involves a chemical reaction between cellulose, sodium monochloroacetate, and caustic. This substance modifies the flow properties of organic and inorganic solutions, such as those found in oil drilling, pharmaceuticals, food production, and detergents.

The production of hydroxyethyl cellulose, carboxymethyl hydroxyethyl cellulose, and methyl cellulose involves caustic. Several additional uses for caustic are outlined (Patel, 2003).

2.3.2. Inorganic Chemicals

The primary application of caustic soda is in producing diverse sodium-based compounds and the sodium salts of acids. Additionally, it is utilized in various other contexts where sustaining a high pH level is necessary. The principal application of sodium hydroxide in the realm of inorganic chemistry is the synthesis of sodium hypochlorite, which finds extensive utilization in domestic and industrial settings for bleaching. The pulp and paper industry employs various sodium-containing compounds, such as sodium hydrosulfide, sodium hydrosulfite (utilized as a bleaching agent for mechanical pulp), sodium sulfide, and sodium sulfite, which also serves as a precursor for the production of sodium thiosulfate (Yadav et al., 2017). Sodium cyanide, which is produced through caustic and hydrogen cyanide reactions, finds application in electroplating and chemical synthesis. Its primary usage, however, is in extracting gold from ores. Zeolites, hydrated alumino silicates containing alkali and alkaline earth metals, have many applications, such as detergent builders, catalysts, and adsorbents. The production of zeolites can be achieved through two distinct methods: the hydrogel process, which entails the gradual crystallization of an amorphous gel resulting from the reaction between sodium aluminate and sodium silicate, leading to the formation of a well-organized structure consisting of sodium cations and aluminate and silicate anions; or the kaolin conversion process, which involves the reaction of acidic with kaolin derived from clay.

Non-crystalline solid solutions known as sodium silicates are commonly utilized in various applications such as adhesives, cleaners, catalysts, and detergents. These solutions are produced through the reaction of acidic or dry soda ash with silica. Sodium phosphates, synthesized through the reaction between phosphoric acid and caustic, find application in the detergents and cleaning agents domain. The utilization of caustic is observed in various food processing procedures, such as eliminating skins from produce items like potatoes and tomatoes (Baena et al., 2023).

2.4. END USES OF POTASSIUM HYDROXIDE (KOH)

Like sodium analog, KOH is a multifaceted intermediary in the chemical manufacturing process. Due to its higher cost, KOH is primarily limited to specialized markets, while NaOH is the dominant choice for large-scale applications. As compared to their sodium counterparts, the increased solubility and reactivity exhibited by certain chemicals in their potassium form are a significant contributing factor to their numerous practical applications. Acetate, bicarbonate, bromide, cyanide, ferrocyanide, oxalate, permanganate, and phosphate are among the commonly utilized derivatives. Potassium carbonate is likely the derivative with the highest market demand, as evidenced by its widespread discussion in academic literature (Al Hamdi & Al Rahmani, 2012).

The primary application of KOH is in manufacturing soaps and detergents. Despite the cost disadvantage, KOH remains a viable option for specialty products, particularly in producing liquid soaps, detergents, and shampoos. KOH finds application in various industries such as glassmaking, petroleum refineries for eliminating sulfides and mercaptans, fertilizers, and numerous other products (Kose et al., 2013).

The global production capacity of KOH is approximately 1.7×10^6 tons per year. It is anticipated that there will be an annual increase in production of approximately 4–5% from the 1.4 × 106 tons manufactured in 2001.

The surplus capacity is expected to diminish in the coming years. However, the extent of readily convertible "swing" capacity from NaOH to KOH production remains uncertain (Goje et al., 2004).

2.5. SUMMARY

Chlorine and caustic soda are closely related chemicals that are commonly produced together through the electrolysis of salt water in the chlor-alkali process. Chlorine has numerous industrial applications, including water treatment, chemical production, pharmaceuticals, paper, and pulp industry. Sodium hydroxide has diverse applications, including chemical manufacturing, textile industry, chemical manufacturing and water treatment. The uses of chlorine and caustic soda can vary depending on specific industries and processes.

REFERENCES

1. Alexander, M., (1985). Biodegradation of organic chemicals. *Environmental Science and Technology, 19*(2), 106–111.
2. Al-Hamdi, K. I., & Al-Rahmani, M. A., (2012). Evaluation of topical potassium hydroxide solution for treatment of plane warts. *Indian Journal of Dermatology, 57*(1), 38.
3. Arévalo-Silva, C., Eliashar, R., Wohlgelernter, J., Elidan, J., & Gross, M., (2006). Ingestion of caustic substances: a 15-year experience. *The Laryngoscope, 116*(8), 1422–1426.
4. Ateia, M., Skala, L. P., Yang, A., & Dichtel, W. R., (2021). Product analysis and insight into the mechanochemical destruction of anionic PFAS with potassium hydroxide. *Journal of Hazardous Materials Advances, 3*, 100014.
5. Baena-Moreno, F. M., Leventaki, E., Ho, P. H., Tajik, A. R., Brzic, D., Sardina, G., & Bernin, D., (2023). Potential of organic carbonates production for efficient carbon dioxide capture, transport, and storage: Reaction performance with sodium hydroxide–ethanol mixtures. *Heliyon, 9*(3), 1–20.
6. Caldwell, D. L., (1981). Production of chlorine. *Comprehensive Treatise of Electrochemistry: Electrochemical Processing*, 105–166.
7. Calvet, R., (1989). Adsorption of organic chemicals in soils. *Environmental Health Perspectives, 83*, 145–177.
8. Camargo, C. L. D. A., Belda Junior, W., Fagundes, L. J., & Romiti, R., (2014). A prospective, open, comparative study of 5% potassium hydroxide solution versus cryotherapy in the treatment of genital warts in men. *Anais Brasileiros de Dermatologia, 89*, 236–240.
9. Elnabarawy, M. T., Welter, A. N., & Robideau, R. R., (1986). Relative sensitivity of three daphnid species to selected organic and inorganic chemicals. *Environmental Toxicology and Chemistry: An International Journal, 5*(4), 393–398.
10. Evans, R. B., (2005). Chlorine: State of the art. *Lung, 183*(3), 151–167.
11. Fawell, J. K., (1993). The impact of inorganic chemicals on water quality and health. *Annali-Istituto Superiore di Sanita, 29*, 293–293.
12. Gao, C., Govind, R., & Tabak, H. H., (1992). Application of the group contribution method for predicting the toxicity of organic chemicals. *Environmental Toxicology and Chemistry: An International Journal, 11*(5), 631–636.
13. Ghannadzadeh, A., & Sadeqzadeh, M., (2017). Exergy aided pinch analysis

to enhance energy integration towards environmental sustainability in a chlorine-caustic soda production process. *Applied Thermal Engineering, 125*, 1518–1529.

14. Giner-Soriano, M., Teixidó, C., Marsal, J. R., Díez, O., Pera, H., Vlacho, B., & working group for the clinical trial IJG-KOH-2014., (2019). Randomized placebo-controlled clinical trial on efficacy and safety of topical 10% potassium hydroxide for molluscum contagiosum treatment in children. *Journal of Dermatological Treatment, 30*(8), 750–756.
15. Gobas, F. A., (1993). A model for predicting the bioaccumulation of hydrophobic organic chemicals in aquatic food-webs: application to Lake Ontario. *Ecological Modelling, 69*(1–2), 1–17.
16. Godlewska-Żyłkiewicz, B., (2006). Microorganisms in inorganic chemical analysis. *Analytical and Bioanalytical Chemistry, 384*, 114–123.
17. Goje, A. S., Thakur, S. A., Diware, V. R., Chauhan, Y. P., & Mishra, S., (2004). Chemical recycling, kinetics, and thermodynamics of hydrolysis of poly(ethylene terephthalate) waste with nonaqueous potassium hydroxide solution. *Polymer-Plastics Technology and Engineering, 43*(2), 369–388.
18. Harrison, E. Z., Oakes, S. R., Hysell, M., & Hay, A., (2006). Organic chemicals in sewage sludges. *Science of the Total Environment, 367*(2–3), 481–497.
19. Harrison, E. Z., Oakes, S. R., Hysell, M., & Hay, A., (2006). Organic chemicals in sewage sludges. *Science of the Total Environment, 367*(2–3), 481–497.
20. Hu, X. M., & Xiao, C. F., (2011). Characterization and hydrolyzation of PAN by sodium hydroxide. *In: Advanced Materials Research* (Vol. 332, pp. 1032–1035).
21. Huckins, J. N., Petty, J. D., & Booij, K., (2006). *Monitors of Organic Chemicals in the Environment: Semipermeable Membrane Devices* (Vol. 223, pp. 1–20).
22. King, F., & Hancock, F. E., (1996). Catalysis and pollution abatement: the removal of hypochlorite from waste chlorine/caustic effluent. *Catalysis Today, 27*(1–2), 203–207.
23. Köse, O., Özmen, İ., & Arca, E., (2013). An open, comparative study of 10% potassium hydroxide solution versus salicylic and lactic acid combination in the treatment of molluscum contagiosum in children. *Journal of Dermatological Treatment, 24*(4), 300–304.
24. Kucek, K. T., César-Oliveira, M. A. F., Wilhelm, H. M., & Ramos, L. P., (2007). Ethanolysis of refined soybean oil assisted by sodium and

potassium hydroxides. *Journal of the American Oil Chemists' Society, 84*(4), 385–392.

25. Kumar, A., Sant, G., Patapy, C., Gianocca, C., & Scrivener, K. L., (2012). The influence of sodium and potassium hydroxide on alite hydration: Experiments and simulations. *Cement and Concrete Research, 42*(11), 1513–1523.
26. Mackay, D., & Paterson, S., (1991). Evaluating the multimedia fate of organic chemicals: A level III fugacity model. *Environmental Science and Technology, 25*(3), 427–436.
27. Martin, W. H., (1961). Potential Competition AND the United States chlorine-alkali industry. *The Journal of Industrial Economics*, 233–247.
28. Muzaffar, F., & Faiz, F., (2014). Comparison of 5% potassium hydroxide with 10% potassium hydroxide solution in treatment of molluscum contagiosum: A comparative study. *Journal of Pakistan Association of Dermatologists, 24*(4), 337–341.
29. Nikravesh, M., Farell, A. E., Lee, C. T., & Van Zee, J. W., (1995). Dynamic matrix control of diaphragm-type chlorine/caustic electrolysers. *Journal of Process Control, 5*(3), 131–136.
30. Pan, B., & Xing, B., (2008). Adsorption mechanisms of organic chemicals on carbon nanotubes. *Environmental Science and Technology, 42*(24), 9005–9013.
31. Patel, M., (2003). Cumulative energy demand (CED) and cumulative CO_2 emissions for products of the organic chemical industry. *Energy, 28*(7), 721–740.
32. Phasey, J., Vandamme, D., & Fallowfield, H. J., (2017). Harvesting of algae in municipal wastewater treatment by calcium phosphate precipitation mediated by photosynthesis, sodium hydroxide and lime. *Algal Research, 27*, 115–120.
33. Plato, C., & Glasgow, A. R., (1969). Differential scanning calorimetry as a general method for determining the purity and heat of fusion of high-purity organic chemicals. Application to 95 compounds. *Analytical Chemistry, 41*(2), 330–336.
34. Riffat, F., & Cheng, A., (2009). Pediatric caustic ingestion: 50 consecutive cases and a review of the literature. *Diseases of the Esophagus, 22*(1), 89–94.
35. Sellers Jr, T., & Prewitt, M. L., (1990). Applications of gel filtration chromatography for resole phenolic resins using aqueous sodium hydroxide as solvent. *Journal of Chromatography A, 513*, 271–278.

36. Sharma, C., Mahanty, S., Kumar, S., & Rao, N. J., (1997). Gas chromatographic determination of pollutants in the chlorination and caustic extraction stage effluent from the bleaching of a bamboo pulp. *Talanta, 44*(10), 1911–1918.
37. Sharma, C., Mohanty, S., Kumar, S., & Rao, N. J., (1996). Gas chromatographic analysis of chlorophenolic, resin, and fatty acids in chlorination and caustic extraction stage effluent from Kahi-grass. *Analyst, 121*(12), 1963–1967.
38. Stolaroff, J. K., Keith, D. W., & Lowry, G. V., (2008). Carbon dioxide capture from atmospheric air using sodium hydroxide spray. *Environmental Science and Technology, 42*(8), 2728–2735.
39. Thomason, J. L., Nagel, U., Yang, L., & Sáez, E., (2016). Regenerating the strength of thermally recycled glass fibers using hot sodium hydroxide. *Composites Part A: Applied Science and Manufacturing, 87*, 220–227.
40. Travis, A. S., (1990). Perkin's mauve: Ancestor of the organic chemical industry. *Technology and Culture, 31*(1), 51–82.
41. Wania, F., & Mackay, D., (1995). A global distribution model for persistent organic chemicals. *Science of the Total Environment, 160*, 211–232.
42. Yadav, V. D., Jain, R., & Dandekar, P., (2017). Influence of sodium hydroxide in enhancing the surface plasmon resonance of silver nanoparticles. *Materials Research Express, 4*(8), 085015.

CHAPTER

3 Applied Processes and Techniques in Chlor-Alkali Industry

Contents

3.1. INTRODUCTION

This chapter provides a qualitative description of the applied processes and techniques. The chloralkali sector employs breaking a salt solution (comprising either potassium or sodium chloride) with water to produce chlorine along with a caustic solution (consisting of sodium or potassium hydroxide) simultaneously. Hydrogen is generated concomitantly via the chlorine along with the caustic solutions (dos Santos Garcia et al., 2019). A chloralkali manufacturing facility in the industrial setting consists of a sequence of procedures.

The chloralkali electrolysis procedure involves the electrolytic decomposition of a chloride salt solution by applying direct current. The mineral sodium chloride is the predominant substance utilized in the process above, with potassium chloride being employed less frequently (constituting roughly 3–4% of the chlorine manufacturing capacity within the EU-27 along with EFTA nations). Potassium chloride is exclusively utilized to produce KOH owing to its significantly elevated raw material expenses (Ali et al., 2015).

The electrolytic manufacturing of chlorine involves three fundamental methods. The cathodic reaction's characteristics are contingent upon the particular methodology employed. The three methodologies encompass the diaphragm technique, which was developed by the Griesheim cell in 1885, the mercury technique, along with the membrane-based cell technique, which emerged in 1970 (Barros et al., 2015). The methodologies exhibit variations within the electrode reactions along with how the resultant chlorine, along with caustic/hydrogen, are segregated.

Did You Know?

Chlorine is a commonly utilized disinfectant in various settings, including pools, purification facilities, and even the home.

Figure 3.1 depicts a simplified diagram of each of the electrolysis process cells.

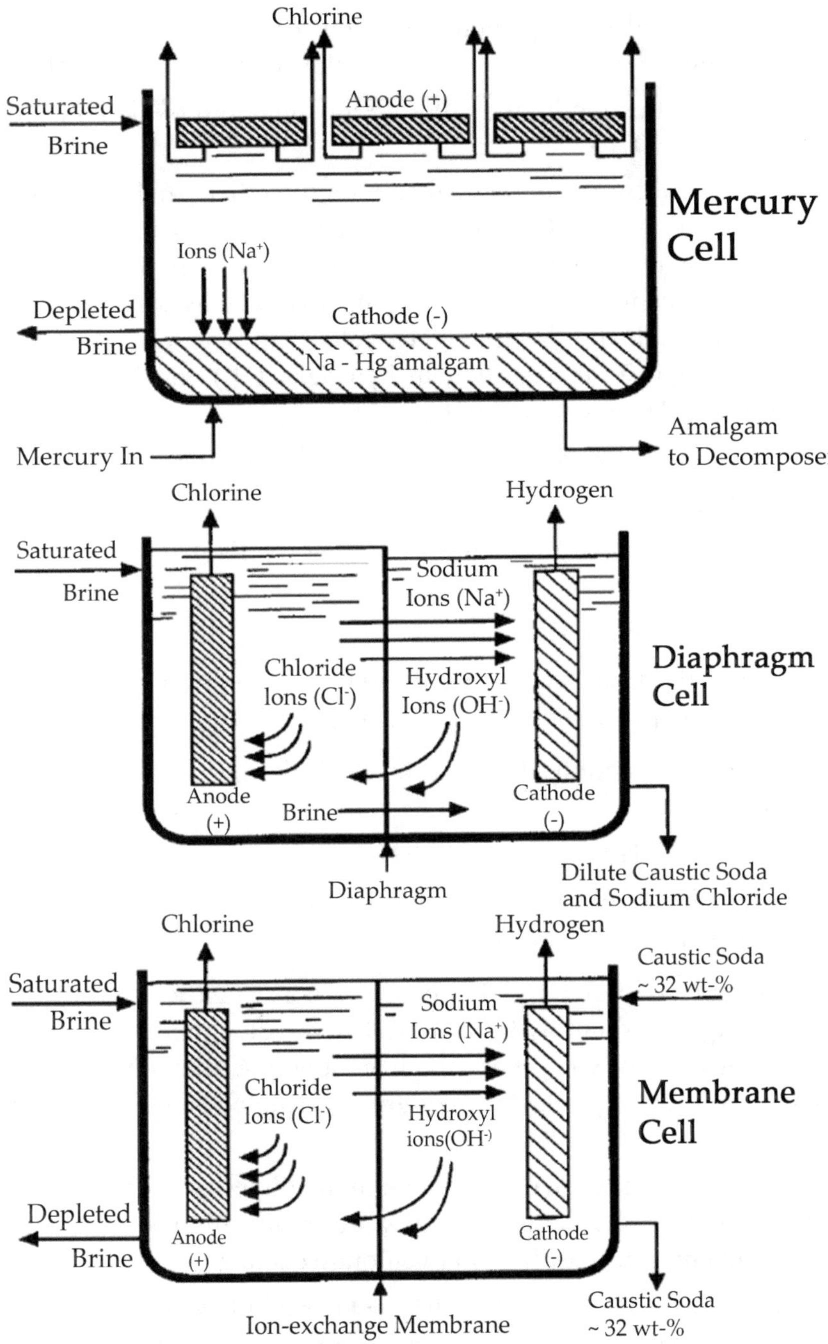

Figure 3.1. *Schematic view of chlorine electrolysis cells.*

Source: Kirk-Othmer, Creative Commons License.

The fundamental concept underlying the process of electrolysis of a solution containing sodium chloride is as follows (Huang et al., 2016):

- Within the anode, the chloride ions undergo oxidation, forming chlorine (Cl_2).
- At a mercury cell, the cathode, on the other hand, facilitates the formation of sodium/mercury amalgams. Afterward, the sodium within the amalgam reacts against water on the decomposer component to produce hydrogen, or H_2, and hydroxides (OH^-). In membrane and diaphragmatic cells, the cathode facilitates water decomposition, resulting in hydrogen (H_2) and hydroxide ions (OH^-) (Huang et al., 2016).

The anode's reaction is consistent across all techniques:

$$2\ Cl^- \rightarrow Cl_2 + 2\ e^-$$

The cathodic process occurring in mercury cells can be described as follows:

$$Na^+ + e^- + Hg_x \rightarrow Na\text{-}Hg_x$$

The chemical process occurring within the decomposer is referred to as:

$$2\ Na\text{-}Hg_x + 2\ H_2O \rightarrow 2\ NaOH + H_2 + 2\ Hg_x$$

The cathodic reaction that occurs in both membranes along with diaphragm cells is:

$$2\ Na^+ + 2\ e^- + 2\ H_2O \rightarrow 2\ NaOH + H_2$$

The comprehensive reaction for all methodologies is:

$$2\ NaCl + 2\ H_2O \rightarrow 2\ NaOH + H_2 + Cl_2$$

The electrolysis process yields products consistently, with a range of 1,070–1,128 kg of sodium hydroxide (100 wt.%) along with roughly 28 kg in H_2 per metric ton of Cl_2 generated (Jeanmonod & Diethelm, 2021). The amalgamation of these products is commonly denoted as the electrical unit (ECU).

Electrolysis may experience a reduction in efficiency due to certain side reactions. At the anode, the process of water oxidization to oxygen along with acid hypochlorous oxidization to chlorate occurs:

$$2\ H_2O \rightarrow O_2 + 4\ H^+ + 4\ e^- \text{ or } 4\ OH^- \rightarrow O_2 + 2\ H_2O + 4\ e^-$$

$$12\ HClO + 6\ H_2O \rightarrow 4\ ClO^- + 8\ Cl^- + 24\ H^+ + 3\ O + 12\ e^-$$

Hypochlorous acid is generated through the disproportionation, also known as dismutation, of chlorine when it is present in water:

$Cl_2 + H_2O \rightleftharpoons HClO + H^+ + Cl^-$

Chlorate is also generated through chemical reactions occurring in the anolyte:

$2\ HClO + ClO^- \rightarrow ClO^- + 2\ Cl^- + 2\ H^+$

Lowering the pH level value is an effective means of suppressing these four significant side reactions. Table 3.1 presents the primary attributes of all three electrolysis methodologies (Zaviska et al., 2012).

Table 3.1. Main Typical Characteristics of the Different Electrolysis Techniques

Criterion	Mercury	Diaphragm	Membrane
Anode	$RuO_2 + TiO_2$ coating on Ti substrate.	$RuO_2 + TiO_2 + SnO_2$ coating on Ti substrate.	$RuO_2 + IrO_2 + TiO_2$ coating on Ti substrate.
Cathode	Mercury	Steel (or steel coated with activated nickel).	Nickel coated with high area nickel-based or noble metal-based coatings.
Separator	None	Asbestos, polymer-modified asbestos, or non-asbestos diaphragm.	Ion-exchange membrane.
Cell voltage	3.15–4.80 V	2.90–3.60 V	2.35–4.00 V
Current density	2.2–14.5 kA/m²	0.8–2.7 kA/m²	1.0–6.5 kA/m²
Temperature	Inlet: 50–75°C Outlet: 80–90°C	NI	NI
pH	2–5	2.5–3.5	2–4
Cathode product	Sodium amalgam ($Na\text{-}Hg_x$)	10–12 wt.% NaOH and H_2	30–33 wt.% NaOH and H_2.
Decomposer product	50 wt.% NaOH and H_2	No decomposer needed.	No decomposer needed.
Evaporator product	No evaporation needed.	50 wt.% NaOH	50 wt.% NaOH
Quality of caustic soda (50 wt.% NaOH).	NaCl: ~ 50 mg/kg $NaClO_3$: ~ 5 mg/kg Hg: ~ 0.1 mg/kg.	NaCl: ~ 10 000 mg/kg (15 000–17 000 mg/kg before concentration) $NaClO_3$: ~ 1 000 mg/kg (400–500 mg/kg before concentration).	NaCl: ~ 50 mg/kg $NaClO_3$: ≤ 10–50 mg/kg
Chlorine quality	O_2: 0.1–0.3 vol-% H_2: 0.1–0.5 vol-% N_2: 0.2–0.5 vol-%	O_2: 0.5–2.0 vol-% H_2: 0.1–0.5 vol-% N_2: 1.0–3.0 vol-%	O_2: 0.5–2.0 vol-% H_2: 0.03–0.3 vol-%

Advantages	50 wt.% high-purity caustic directly from cell, high-purity chlorine and hydrogen, simple brine purification.	Low quality requirements of brine, low electrical energy consumption.	Low total energy consumption, low investment and operating costs, no use of mercury or asbestos, high-purity caustic, further improvements expected.
Disadvantages	Use of mercury, expensive cell operation, costly environmental protection, large floor space.	High steam consumption for caustic concentration in expensive multi-effect evaporators, low-purity caustic, low chlorine quality, some cells are operated with asbestos diaphragms.	High-purity brine required, low chlorine quality, high cost of membranes.

Source: Thomas Brinkmann, Creative Commons License.

3.2. THE MERCURY CELL TECHNIQUE

3.2.1. General Description

The utilization of the mercury cell methodology has been prevalent in Europe since the year 1892. The utilization of the mercury cell method involves the incorporation of an electrolysis process cell and a decomposer, which may be either horizontal or vertical, as illustrated in Figure 3.2. The electrolysis process involves using a marginally inclined trough, through which purified and saturated brine containing roughly 25% sodium chloride flows (Gardiner, 1953). A thin layer of mercury (Hg) is observed to flow alongside the brine within the trough's base. An anode assemblage is suspended near the cathode.

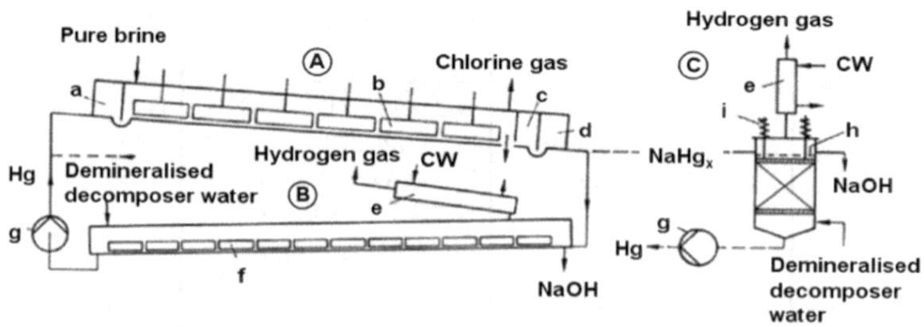

Figure 3.2. *Schematic view of a mercury electrolysis cell with horizontal and vertical decomposer.*

Source: A. Ullmann, Creative Commons License.

The utilization of the mercury cell methodology has been prevalent in Europe since the year 1892. The utilization of the mercury cell method involves the incorporation of an electrolysis process cell and a decomposer, which may be either horizontal or vertical, as illustrated in Figure 3.2. The electrolysis process involves using a marginally inclined trough, through which purified and saturated brine containing roughly 25% sodium chloride flows (Jakšić et al., 1968). A thin layer of mercury (Hg) is observed to flow alongside the brine within the trough's base. An anode assemblage is suspended near the cathode.

The anolyte that exits the cell has been depleted and saturated with chlorine, thus requiring partial dechlorination before reentry into the dissolvers.

Sodium hydroxide is generated through decomposition, yielding a roughly 50% weight concentration (Hocking & Hocking, 1985). The highest concentration recorded is 73 weight percent. Nonetheless, it is noteworthy that no plant within the EU-27, along with EFTA nations, has been identified as currently functioning beyond a weight percentage of 50.

The functionality of the mercury cell is reliant on the elevated excessive potential of the hydrogen upon mercury, which facilitates the selective discharge of sodium over hydrogen. Nonetheless, the presence of impurities, namely vanadium (V), chromium (Cr), and molybdenum (Mo), during concentrations ranging from 0.01 to 0.1 parts per million (ppm), as well as other elements such as aluminum (Al), barium (Ba), calcium (Ca), cobalt (Co), iron (Fe), magnesium (Mg), nickel (Ni), and tungsten (W) at the ppm level, can compromise the overpotential protection of the mercury surface, leading to the localized discharge of hydrogen into the chlorine (Lyu et al., 2024). The possibility exists that the concentration of hydrogen within the chlorine may escalate to a level where the cell and subsequent equipment for handling chlorine could harbor combustible amalgamations.

Typically, the operation of mercury cells involves the maintenance of a concentration of salt within the spent brine released from the cell, which ranges from 21 to 22 weight percent. This indicates the disintegration of approximately 15–16% of the salt in a solitary iteration. Reducing salt concentration through additional decomposition within the brine would result in a reduction in brine conductivity, leading to a reduction in electrical efficiency. Nevertheless, it has been observed that in plants utilizing once-through brine systems, around 40% of the salt undergoes electrolysis within the cells. Compared to the diaphragm and membrane cell techniques, the element mercury cell technique generates chlorine gas containing minimal amounts of oxygen and a 50-weight percent caustic soda solution (Ferrara et al., 1992). Mercury cells, in contrast to the diaphragm or membrane cells, function at a greater voltage (3–5 V) along with current density (7–10 kA/m^2), resulting in higher energy consumption (excluding caustic soda concentration). The methodology necessitates a brine solution that is pure and devoid of metal impurities to mitigate the potential hazard of hydrogen production within the cell, leading to an explosion. The utilization of the mercury cell procedure results in the inevitable occurrence of environmental discharges of mercury.

Did You Know?

The Chloralkali company is widely recognized as a prominent and enduring electrochemical sector globally, boasting a rich legacy that can be traced back to the 1800s. The chloralkali sector is widely recognized as a prominent and enduring electrochemical industry on a global scale, boasting a rich legacy that traces back to the 1800s.

3.2.2. The Cell and Decomposer

The cellular structure comprises a prolonged, marginally inclined channel with a slope range of 1.0–2.5% and a hermetically sealed lid (Rath & Rousk, 2015). The trough is composed of steel material and is enveloped by a non-conductive rubbery coating that prevents any interaction with the anolyte. Additionally, it confines the brine-cathode connection with the mercury exterior and prevents the corrosion of the electrolyte. The dimensions of cells are typically within the range of 1 to 2.5 meters in width and 10 to 25 meters in length. Consequently, the cellular region has the potential to exceed 30 square meters. The cellular dimensions may exhibit considerable variability to achieve the targeted rate of chlorine generation (Southworth et al., 2004). The steel substrate is engineered to possess a high degree of surface smoothness, thereby facilitating the continuous flow of mercury in an uninterrupted film. If a rupture occurs on the surface of mercury, it will form caustic soda on the exposed cathode made of steel. This process will occur concurrently with the liberation of hydrogen, which will subsequently combine with chlorine. The cathode is constructed with a thin layer of mercury that exhibits a unidirectional flow across the cell, facilitated by the slight tilt of the cell base. The mercury contained within a single cell may reach a maximum of 6 metric tons. A high level of pureness, $\geq$ 99.9 wt.% of the element mercury, and very low levels of metals, within particularly heaviest metals, are required. Typically, metallic mercury not utilized in electrolysis is stored within a distinct storage location, enclosed within bottles or containers (Sanders et al., 1953). For over six decades, graphite was the sole material employed as the anode for chlorine manufacturing. However, despite its usage, it displayed a high chlorine excessive potential and dimensional instability due to the electrochemical conversion of carbon with carbon dioxide. This resulted in elevated energy consumption and necessitated regular upkeep operations. During the latter part of the 1960s, anodes were made of titanium coated via ruthenium dioxide (RuO_2) and titanium dioxide (TiO_2) (Kinsey et al., 2004). These anodes were referred to as dimensionally resistant anodes. The utilization of metal anodes coated with RuO_2 and TiO_2 has resulted in a reduction in energy consumption of roughly 10% compared to graphite.

Furthermore, the lifespan of these coated anodes is higher, with a production capacity of 300–400 t Cl_2/m^2. Their lifespan varies from less than a year to over five years, according to the current density and the distance between the anode and cathode. The optimization of anode geometry within mercury cells is currently undertaken to enhance gas release, thereby mitigating ohmic losses and augmenting the uniformity of the brine, which in turn improves the longevity of anode coating (De Nora, 1950).

Figure 3.3. *View of a mercury cell room.*

Source: BASF, Creative Commons License.

A cellular structure has an 'end-box' component affixed to both extremities. The terminal apparatus features distinct sections for gathering chlorine gas and weirs for segregating the mercury or brine flows. Additionally, it enables the cleansing of the mercury and facilitates the elimination of the dense mercury 'butter' that arises due to impurities.

To prevent the occurrence of stray ground currents, the entire cell has been insulated from the floor. Typically, multiple cells are arranged in a series configuration through the electrical interconnection of the cathode for one cell with the anodes for the next one (Gardiner, 1978). It is possible to repair and substitute individual cells by bypassing them. Typically, cells are located within a structure, although occasionally, they may be constructed in an outdoor setting (Figure 3.3).

Typically, the electrolysis cell is subjected to brine inflow at a temperature range of 60–70°C. The brine solution exhibits more conductivity at the given temperature, and the mercury displays greater fluidity than under ambient conditions (Landis et al., 2004). The attainment of temperature can be facilitated through the preheating of the saturated brine, followed by its elevation within the cell through the process of Joule heating, ultimately reaching a range of 75–85°C. The thermal energy generated during the processes of the electrolysis process and amalgam decay necessitates the replacement of the ambient air within the cell room at a frequency ranging from 10 to 25 times each hour, contingent upon the specific characteristics of the edifice (Lakshmanan & Murugesan, 2014).

Decomposing the amalgam typically involves horizontal decomposers, either next to or beneath the cell. However, it has become more prevalent since the 1960s to employ vertical decomposers, which are located at one end of the cell. Industrial decomposers can be described as primary electrochemical cells that have been short-circuited. The cathode of these cells comprises a graphite catalyst, while the anode is composed of sodium amalgam (Reis et al., 2009). Graphite is the prevailing catalyst, typically activated by iron, nickel, or cobalt, which oxides or through molybdenum or tungsten-based carbides. The temperature range of the decomposer is estimated to be between 90°C and 130°C, resulting from both the chemical reactions occurring within the decomposing agent and the infusion of warm amalgam to use coming from the cell. Elevated temperatures reduce hydrogen overpotential on graphite, accelerating the reaction rate (Gibičar et al., 2009).

3.3. THE DIAPHRAGM CELL TECHNIQUE

3.3.1. General Description

The diaphragm cell methodology originated during the 1880s in the US and constituted the initial commercial method employed for the production of chlorine along with caustic soda via brine. The methodology employed in this technique is distinct from that of the mercury cell technique, as all chemical reactions occur within a single cell, producing a cell effluent comprising bath salts that have been dissolved along with caustic soda (Stokes, 1950). A diaphragm segregates the chlorine released at the anode from the molecules of hydrogen and acidic soda generated directly during the cathode. Without a diaphragm, the coexistence of hydrogen and chlorine could end in spontaneous ignition. Additionally, the reaction between caustic soda and chlorine would yield sodium hypochlorite (NaClO), which would further respond to produce sodium chlorate ($NaClO_3$).

The diaphragm facilitates the separation of the feed brine, also known as anolyte, from the catholyte, which includes caustic. The refined saline solution is introduced into the anodic section and permeates through the partition into the cathodic region. The regulation of the percolation rate is achieved by maintaining an elevated liquid level within the anode compartment, which establishes a meticulously controlled and affirmative hydrostatic head (Easteal et al., 1989). The regulation of the diffusion rate is crucial in achieving a suitable equilibrium between a reduced rate, resulting in a desirable level of caustic soda within the catholyte (the effluent of the cell), and an increased rate, which serves to restrict the reverse movement of hydroxyl ions within the catholyte through the anolyte, thereby mitigating the reduction in cathode current efficiency.

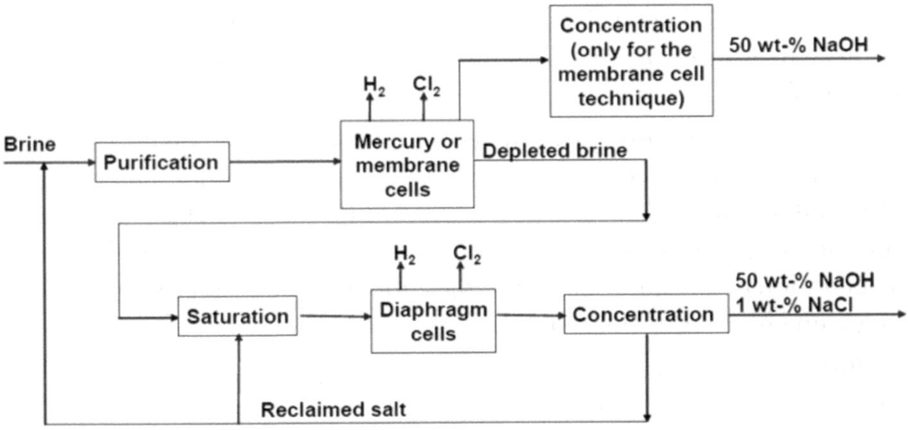

Figure 3.4. *Flow diagram of integrating the membrane or mercury and the diaphragm cell techniques.*

Source: Serge Roudier, Creative Commons License.

The diaphragm cells typically generate a cell liquor that comprises approximately 10–12 weight percent of sodium hydroxide and 15–17 weight percent of sodium chloride. Typically, this solution is concentrated to a 50-weight percent of sodium hydroxide. If utilized without additional concentration, the resulting caustic liquor may comprise reduced levels of roughly 7-weight percent sodium chloride. Most sodium chloride undergoes precipitation in evaporation, leaving approximately 1.0 weight percent residual. The salt produced exhibits high purity and is commonly employed to generate additional brine. This sodium chloride of superior quality is occasionally employed as a primary substance for mercury as a component or membrane cell. Figure 3.4 depicts a potential integrated plant as a flow diagram (Little et al., 1991).

The electrolytic decay of water results in a small amount of oxygen (0.5–2.0 vol-%) in chlorine. In addition, it has been observed that chlorate is generated within the cell liquor through anodic oxidation and the dispersion of hypochlorous acid. Before concentration, the total amount of hypochlorous acid is approximately 0.04–0.05 wt.%, which increases to approximately 0.1 wt.% following concentration (Cussler Jr & Dunlop, 1966).

The diaphragm cell facilitates the decomposition of both saturated brines, containing roughly 25 weight percent of NaCl, resulting in a reduction of approximately 50% of its initial concentration upon passing through the cell. This is in contrast to the mercury cells, which exhibit a decay rate of 16% per passage. Applying an electrical current to liquid results in the generation of heat, which in turn elevates the temperature within the electrolyte to a range of 80–99°C.

Diaphragm cells offer certain benefits, such as low-quality demands for the brine and low electrical energy consumption, with a cell voltage of 3–4 V and a current density of 0.5–3 kA/m^2 (Stokes, 1951). Nevertheless, a substantial quantity of steam might be requisite for achieving the desired concentration of caustic soda, while the caliber of the resultant caustic soda, along with chlorine, is suboptimal.

Utilizing asbestos diaphragms within the diaphragm cell method results in the inherent emission of asbestos.

3.3.2. The Cell

Several diaphragm cell designs are being devised and implemented for industrial purposes. The schematic representation of a standard monopolar diaphragm cell is depicted in Figure 3.5, while Figure 3.6 illustrates a monopolar diaphragmatic cell chamber. The customary range of anode regions per cell is between 20 and 100 square meters.

The cathodes employed in diaphragm cells are typically composed of carbon steel and are coated with an active substance that reduces hydrogen overpotential (Krol et al., 1991). This results in notable reductions in energy consumption. The coatings are comprised of at least two constituents. An extremely porous nickel exterior is obtained by subjecting at least one component to caustic leaching. The coatings must exhibit robustness, given that a high-pressure water jet is employed to extract the diaphragm via the cathode mesh upon terminating its lifespan. This process has the potential to impact the coatings negatively.

The anodes utilized in diaphragm cells comprise titanium coated with ruthenium dioxide, tin dioxide, and titanium dioxide (Belhadj et al., 2022). The expandable anode design is widely accepted in commercial applications because it can compress the anode framework during cell assemblage and expand while the cathode is in place. The cathode spacers are strategically positioned to establish a regulated distance of several millimeters between the anode and the cathode. The reduction of the gap results in a decrease in electricity use.

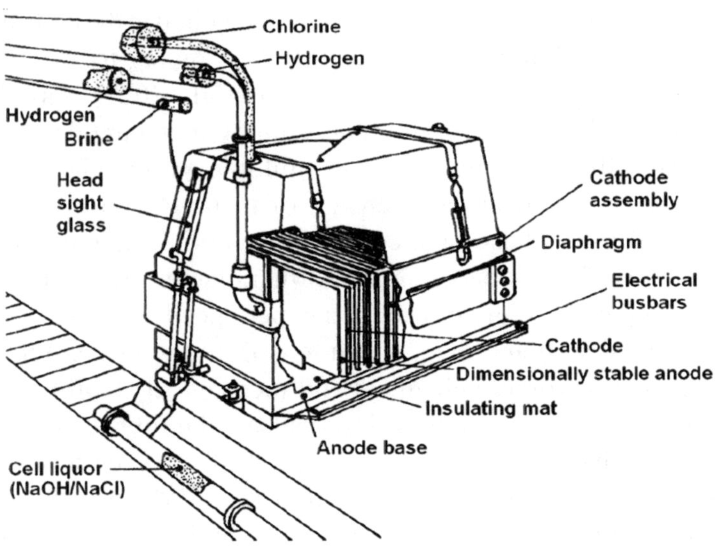

Figure 3.5. *Schematic view of a typical monopolar diaphragm cell.*

Source: Le Chlore, Creative Commons License.

Figure 3.6. *View an open-air diaphragm cell room equipped with monopolar electrolyzers.*

Source: Tilak V. Bommaraju, Creative Commons License.

In the past, the utilization of graphite anodes resulted in the malfunctioning of the diaphragm within a span of 90–100 days owing to the obstruction caused by graphite particulates (Abd-Almohi et al., 2022). As of the year 2011, it has been observed that all plants employ metal anodes, along with the diaphragm's lifespan can extend to multiple years. The alteration in composition has resulted in an extension of their service life. During the late 1920s, the deposited asbestos-containing materials diaphragm replaced the initial use of asbestos paper sheets in manufacturing diaphragms. The diaphragms composed solely of asbestos experienced swift obstruction due to the influx of calcium and magnesium ions from the brine. Asbestos was selected due to its favorable chemical and mechanical stability, as well as its comparatively low cost and widespread availability. Commencing in the early 1970s, diaphragms solely composed of asbestos were gradually substituted with diaphragms consisting of at least 75% asbestos and up to 25% fibrous fluorocarbon polymer, which exhibited superior resistance to chemicals (Dötzel & Schneider, 2002). The PMA diaphragms exhibit enhanced stability due to incorporating polymers that stabilize the asbestos. This results in a reduction of the cell voltage and facilitates the utilization of an expandable anode. The exclusive type of asbestos utilized in diaphragm cells was chrysotile, also known as 'white asbestos.

Efforts are being undertaken to substitute asbestos with alternative diaphragm materials because of the potential for personnel exposure to asbestos and the consequent environmental emissions.

The inception of non-asbestos diaphragms commenced during the mid-1980s. In 2010, two readily accessible diaphragm systems were asbestos-free (Nawrat et al., 2003). Each asbestos-free diaphragms share a common foundation material, primarily composed of a fluorocarbon polymer, specifically PTFE (polytetrafluoroethylene). The variations are discernible in the types of fillers employed and the techniques applied to the hydrophobic PTFE fibers to create a permeable and hydrophilic abdominal wall.

3.4. THE MEMBRANE CELL TECHNIQUE

3.4.1. General Description

In the early 1970s, the advent of ion-exchange membranes marked the inception of a novel approach to the production of chlorine, known as the membrane-based cell technique (Crook & Mousavi, 2016). In 1975, the inaugural industrial membranes cell plant was installed in Japan. As a result of the stringent environmental regulations imposed by the Japanese government following the Minamata disease outbreak, which was attributed to the discharge of methylmercury-contaminated wastewater into Minamata Bay during the 1950s, Japan became the pioneer in the widespread adoption of the technique in the mid-1980s. Because the 1990s, the membrane-based cell technique has been regarded as the cutting-edge method for manufacturing chlorine and caustic soda/potash (Snyder et al., 2003).

The present methodology involves the implementation of an ion-permeable membrane to segregate the anode and cathode. The anode compartment is traversed by a brine solution, wherein chloride ions are oxidized to chlorine gas. The migration of sodium ions, accompanied by a ratio of 3.5 to 4.5 moles of liquid per mole of sodium, occurs across the membrane towards the cathode compartment, where a solution for caustic soda is present. The gases hydrogen and hydroxide ions are released through the process of electrolysis at the cathode. The amalgamation of sodium along with hydroxide ions results in the formation of caustic soda (Suleman et al., 2016). This compound is usually maintained at 32 ± 1 wt.% within the cell. This is achieved by blending a fraction of the result stream using demineralized drinking water to attain a concentration of roughly 30 wt.%. The resulting mixture is then recycled through the catholyte inlet.

The process involves the continuous extraction of caustic soda coming from the circuit. The anode compartments release depleted brine which is subsequently re-saturated with salt. The membrane effectively impedes the transference of chloride ions compared to the anode chamber to the cathode storage, producing a caustic soda solution with minimal sodium chloride content (around 50 mg/l). The membrane effectively hinders the occurrence of back-migration from hydroxide; however, it occurs to a certain degree and contributes to the generation of oxygen, hypochlorite, and chlorate within the anode compartment (Garcia-Herrero et al., 2017). This, in turn, reduces the present level of caustic soda manufacturing by 3–7%.

Certain electrolyzers generate a comparatively less concentrated caustic soda solution, with a weight percentage between 23 and 24. The concentration within the caustic solution introduced into the cell is estimated at around 20–21 wt.%. The energy released during the electrolysis process can be harnessed to increase the concentration of the caustic solution from 23–24 wt.% to 32–34 wt.%. The energy efficiency of the process in question is similar to that of the previous method that utilizes a 32 wt.% caustic solution. However, the former process necessitates additional equipment for caustic evaporation (Banerjee & Curtin, 2004). Conversely, less complex and affordable building materials may be employed in the corrosive pathway encircling the membrane units.

Typically, the caustic solution with a concentration ranging from 30–33 wt.% becomes concentrated to the typical commercial level of 50 wt.% through evaporation, which involves using steam. An alternative approach entails utilizing the caustic generated in the outer layer of cells to nourish the decomposing agents of mercury cells (Erxleben & Ruzicka, 2005) (Figure 3.7).

The sodium chlorate level within the resulting caustic soda generally falls between ≤ 10 and 50 mg/kg. The operational current density, membrane features, and chlorate levels within the brine determine the level (Oehmen et al., 2014). The chlorine generated through the process of membrane cells is characterized by relatively low oxygen levels, typically ranging from 0.5 to 2.0 volume percent. Depression of oxygen and chlorate creation can be achieved by choosing an anode coating with appropriate characteristics and reducing the anode area's pH level.

The loss of brine in membrane cells is observed to be two to three times higher than that in mercury cells. This characteristic enables the brine systems to become more compact, reducing recycling rates and equipment requirements compared to mercury cell facilities with equivalent capacity (Divisek et al., 1998).

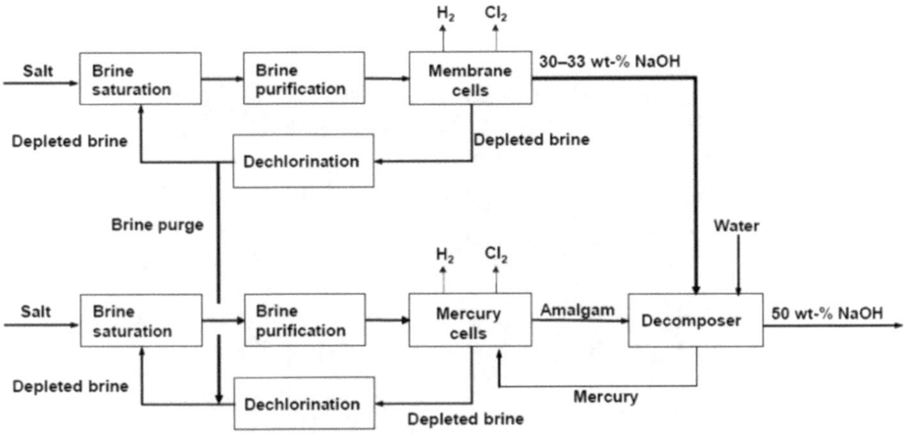

Figure 3.7. *Flow diagram of the integration of the membrane and mercury cell techniques.*

Source: Frauke Schorcht, Creative Commons License.

Utilizing the membrane-based cell technique generates a highly refined caustic soda solution while consuming comparatively lower energy than alternative methodologies. Furthermore, the membrane cell method does not involve mercury, categorized as highly toxic, nor asbestos, regarded as a toxic substance with carcinogenic properties. The membrane-based cell method presents certain drawbacks, such as the potential requirement to increase the amount of caustic soda manufactured through evaporation (Oehmen et al., 2006). Additionally, in certain programs, the chlorine gas generated may necessitate processing to eliminate oxygen, typically accomplished through liquefaction and subsequent evaporation. In addition, the brine introduced into a membrane cell must exhibit a high degree of purity. This necessitates the implementation of supplementary purification procedures before the commencement of electrolysis.

3.4.2. The Cell

Several membrane cell designs have already been devised and implemented in commercial applications. The schematic representation of a standard bipolar membrane in the electrolysis process cell is depicted in Figure 3.8. Additionally, Figure 3.9 illustrates a practical instance of a bipolar membrane cell room, while Figure 3.10 displays an example of a monopolar membrane cell room (Li et al., 2018).

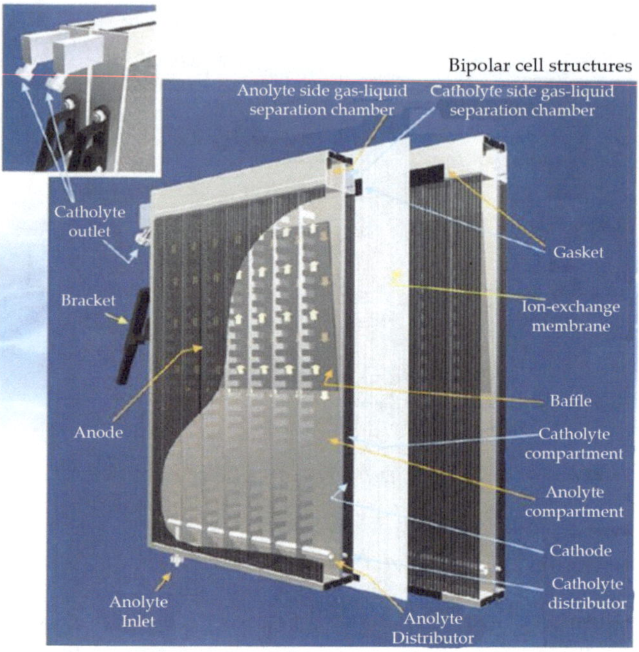

Figure 3.8. *Schematic view of a typical bipolar membrane electrolysis cell.*

Source: Asahi Kasei, Creative Commons License.

Figure 3.9. *View a membrane cell room with bipolar electrolyzers.*

Source: Vinnolit, Creative Commons License.

Figure 3.10. *View a membrane cell room equipped with monopolar electrolyzers.*

Source: Solvay, Creative Commons License.

Nickel has become the cathode material employed in membrane cells. Similar to the cathodes found in diaphragm cells, it is common for them to be coated via a catalyst that exhibits greater stability than the substrate (Vargas-Barbosa et al., 2014). This coating serves to enhance the surface area and decrease the overpotential. The materials utilized for coating purposes encompass a range of compositions, such as Ni-S, and Ni-Al, along with Ni-NiO blends and amalgamations of nickel along with platinum group metallic materials. Due to the elevated caustic concentration, cathode coatings utilized in membrane cells must exhibit superior chemical resistance to those employed in diaphragm cells.

The utilized anodes comprise titanium coated with a blend of iridium dioxide, ruthenium dioxide, and titanium dioxide (Li et al., 2016). During the onset of the 21st century, it was observed that the second-generation coatings utilized in membrane cells exhibited lifetimes that were equivalent to those found in the diaphragm cell method.

Perfluorinated polymers are frequently utilized as membranes in the chlor-alkali sector. The composition of membranes typically comprises between one and three layers, with a predominant occurrence of two layers. One stratum comprises a perfluorinated polymer that features replaced carboxylic groups and is positioned near the cathode (Xu et al., 2020). The second stratum comprises a type of perfluorinated polymer that features replaced sulfonic groups and is situated near the anodic surface. The outermost layer of carboxylate demonstrates a notable preference for transporting potassium and sodium ions, effectively impeding the transportation of hydroxide and chloride. In addition, the layer of sulfonate confers favorable mechanical resilience and elevated electrical conductivity. PTFE fibers are commonly used for reinforcement to enhance the mechanical robustness of the membrane. The membranes must maintain stability despite being subjected to chlorine on one side and a potent acidic solution on the other. Membranes available in the market are designed and tailored to cater to a particular caustic concentration for optimal performance. The membrane sizes vary between 0.2 m^2 and 5 m^2, contingent upon the specific design (Marin et al., 2023). The typical operational lifespan for chlor-alkali membranes is estimated to be between three and five years (Figure 3.11).

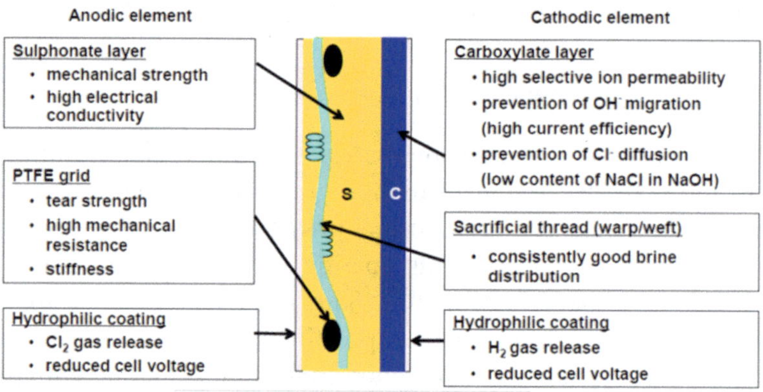

Figure 3.11. *Schematic view of a membrane.*

Source: Asahi Kasei, Creative Commons License.

The reduction of voltage drop throughout the electrolyte within a membrane cell can be accomplished through electrode proximity optimization. In instances where the gap is minimal, the voltage experiences an increase due to the confinement of gas bubbles amidst the electrodes and the hydrophobic membrane (Schneiders & Luke, 1992). To mitigate this phenomenon, a strategy involves applying a porous inorganic substance as a thin coating on each side of the membrane, thereby augmenting its surface's capacity to discharge gaseous byproducts. The enhanced membranes have facilitated the advancement of contemporary cells featuring cathode structures with either zero-gap or finite-gap configurations.

3.5. BRINE SUPPLY

3.5.1. Sources, Qualities, and Storage of Salt

Typically, the brine utilized in mercury and membrane cells is obtained using the dissolution of solid salt within depleted brine. However, certain facilities opt for solution-mined brine, utilized once through, meaning there is no brine recirculation (LeRoy, 1983). The utilization of brine supply within diaphragm cells is typically a single-pass process, predominantly employing solution-mined brine. However, recycling salt retrieved through caustic evaporator units back into the brine supply is plausible.

Typically, the fundamental primary substance utilized is in the form of solid salt. This may include rock salt, which is extracted through mechanical mining processes, solar salt, which is generated through the evaporated water of seawater and brine; or vacuum salt, which is derived from the purification and evaporated water solution-mined brine (Hanson et al., 2004). Additional sources of salt comprise the salt retrieved through the harsh evaporators from the diaphragmatic cell method and the brine acquired through solution mining, which involves the injection of water or weakened brine through a salt deposit to dissolve the substance and transport it to the surface. In Spain, an additional resource is utilized whereby 70% of the total chloride of sodium utilized in chlor-alkali vegetation is acquired through the refinement of NaCl-containing waste materials originating from the mining of potash (KCl), which encompasses waste materials from antiquated landfills. Under certain circumstances, it is feasible to repurpose wastewater containing salt that originates from alternative manufacturing procedures (Gude, 2015).

In 2011, mercury cell crops in the EU-27, along with EFTA countries, predominantly utilized vacuum salt and rock salt. Solution-mined brine was the primary choice for diaphragm cell plants, while membrane cell plants mainly relied on vacuum salt and, to a lesser degree, solution-mined brine and rock salt (Kaplan et al., 2017). A limited number of Southern European plants exclusively utilized solar salt, while those in Spain relied on salt derived from potash mineral extraction wastes. Certain plant species employ a mixture of various types of salts. The chemical compositions derived from diverse sources exhibit significant variations; however, a ubiquitous contaminant in almost all salts is a variant of calcium sulfate. Solar salt typically exhibits higher purity levels than rock salt, particularly following the customary washing procedure. Nevertheless, it is more prone to caking and mechanical deterioration (Ahmad & Baddour, 2014). Typically, rock salt exhibits a higher concentration of calcium sulfate and a relatively lower proportion of magnesium. The enhancement of magnesium hydroxide precipitate can be attributed to the developing hybrid particles facilitated by a higher calcium-to-magnesium ratio. Several chlor-alkali plants function using salt, which has undergone a certain degree of purification. Solar salts are typically subjected to a washing process to eliminate any occluded liquor along with surface impurities (Vanhaecke et al., 1990). On the other hand, vacuum salt is reconstituted from brine after removing most impurities through chemical treatment. The chemical constitution of the salt acquired through the refinement process of potash mining residuals in Spain falls within the compositional spectrum of rock salt, as presented in Table 3.2. Salt obtained through solution mining is anticipated to possess compositions akin to rock salt, albeit with substantially lower amounts of insoluble substances (De Vito et al., 2011).

Table 3.2. Typical Compositions of Sodium Chloride Used in Chlor-Alkali Electrolysis

Component	Salt Source (compositions are given on a dry basis)		
	Rock Salt	**Washed Solar Salt**	**Vacuum Salt**
NaCl	93–99%	99%	99.95%
SO_4^{2-}	0.2–1%	0.2%	0.04%
Ca^{2+}	0.05–0.4%	0.04%	0.001 2%
Mg^{2+}	0.01–0.1%	0.01%	0.000 1%

Source: Thomas Brinkmann, Creative Commons License.

Typically, the salt is kept in a securely enclosed space outfitted with a protective covering to prevent any potential dispersion beyond the designated area. Protective measures are implemented to avert contamination, particularly in precipitation or when the storage facility is near surface fluid or groundwater (Govindarajan & Connell, 1983). Due to its elevated purity level, it is imperative to safeguard vacuum salt. Ferrocyanides, primarily anti-caking agents, are commonly incorporated into vacuum salt to prevent caking. However, their use in rock salt could be more frequent.

Brine obtained through solution mining is occasionally conveyed via pipelines, covering distances that may surpass 100 km. The saline solution is stored in reservoirs.

3.5.2. Brine Preparation

When utilizing solid salt as the primary substance, a dissolution process becomes imperative, which can be executed in vessels that are either open or closed (Altheide et al., 2009). The process of saturating salt can involve applying water and drained brine by either spraying onto the rock salt or introducing it at the bottom of the saturator gradually. In the latter scenario, the brine that has reached its saturation point exceeds the capacity of the equipment and spills over from the upper section. Contemporary saturators are enclosed containers designed to minimize the discharge of saltwater spray and mist, in addition to mercury, in instances where the mercury cell methodology is employed (Shen & Wang, 2011). The total amounts of NaCl found in the water-saturated brine are observed to attain levels ranging from 310 to 315 g/l.

3.5.3. Brine Purification

3.5.3.1. General Description

Purifying brine involves a primary system encompassing the three cell methods, including precipitation and filtration. Additionally, the membrane cell technique requires an auxiliary secondary system (Wang & Li, 2023). Brine purification is a necessary step to decrease the levels of unwanted constituents, such as sulfate anions, cations of calcium, magnesium, barium, and metals, which have the potential to impact the electrolytic process negatively. The level of complexity of a brine therapy unit is determined by the quality of the starting point, along with the brine quality standards associated with every one of the three methods.

3.5.3.2. Primary Purification

In the first phase of the purification process, sodium carbonate, along with sodium hydroxide, is employed to induce the precipitation of calcium and magnesium ions in the form of calcium carbonate ($CaCO_3$) and magnesium hydroxide ($Mg(OH)_2$) (Luo et al., 2021). During this process, certain metals such as iron, titanium, chromium, nickel, molybdenum, and tungsten can undergo precipitation in the form of hydroxides. Typically, the conventional approach for mitigating metal concentrations involves establishing maximum concentration thresholds within the procurement requirements for the salt. The management of sodium sulfate can be achieved through the introduction of calcium chloride, also known as $CaCl_2$, or barium compounds

($BaCO_3$ or $BaCl_2$), which facilitate the elimination of sulfate anions via the formation of sulfate of calcium ($CaSO_4$) along with barium sulfate ($BaSO_4$). The simultaneous precipitate of barium sulfate may occur alongside calcium carbonate and magnesium hydroxide precipitation, though calcium sulfate precipitation necessitates a distinct vessel.

In cases where vacuum salt is utilized as the main raw material, it is common practice to subject only a portion of the brine stream to primary purification treatment. In contrast, the entire stream is typically treated when alternative salt types are employed (Langdon et al., 1990). Certain plant species employ vacuum salt and eliminate the primary brine purification process.

Eliminating precipitated impurities involves either sedimentation, filtration, or a hybrid of both techniques. Typically, the filter cake that has been separated has been concentrated to attain a solids content of 50–60% using presses, rotary drum vacuum filters, or centrifuges before disposal (van Hemert et al., 1984).

The reduction of sulfate content may be achieved through various methods that do not require costly barium salts. These methods include purging a portion of the brine, cooling the water stream and inducing crystallization of $Na_2SO_4 \cdot 10H_2O$, precipitation of the dual salt $Na_2SO_4 \cdot CaSO_4$, ion exchange, or a combination of nanofiltration and brine purging. In the diaphragm cell method, eliminating sulfate may not always be a requisite procedure, as the pure form of Na_2SO_4 can be extracted from the cell liquor during the concentration phase (Ilyas et al., 2021). The avoidance of salts of barium is a common practice in the membrane of the cell technique to safeguard the membrane from possible precipitations.

Ideally, the brine that has undergone purification should contain:

- Calcium (Ca^{2+}): < 2 mg/l;
- Magnesium (Mg^{2+}): < 1 mg/l;
- Sulfate (SO^{2-}): < 5 g/l.

Typically, before its introduction into mercury or diaphragmatic electrolysis cells, the brine undergoes acidification with hydrochloric acid to achieve a pH level below 6. This process extends the lifespan of the anode coating and minimizes the production of oxygen, hypochlorite, and chlorate.

3.5.3.3. Control of Nitrogen Compounds in the Brine

The occurrence of certain nitrogen compounds within the brine results in the generation of nitrogen trichloride, a highly combustible compound (Kundu, 2010).

3.5.4. Brine Dichlorination and Resaturation

Typically, mercury or membrane cell facilities employ the practice of brine recirculation along with resaturation. In 2011, there existed three operational plants that employed a once-through approach. One of these plants utilized the mercury cell method, while another employed the membrane cell process (Khan et al., 2015). The third plant utilized both mercury and membrane cell techniques. The botanical specimens are situated in Spain, Portugal, and the UK near the ocean and substantial subterranean salt reservoirs. Diaphragm cell facilities typically utilize a singular brine circuit, although certain plants may incorporate brine saturation utilizing salt retrieved via caustic evaporators.

The initial step in recirculation circuits involves dichlorination of the brine that has been depleted after passing through the electrolyzes (Dalvi et al., 2000):

- The mercury cell method is only partially effective in preventing the presence of metallic mercury within the brine purification process sludge. This is because departing active chlorine within the brine results in the mercury remaining oxidized, such as $HgCl_3^-$ and $HgCl_4^{2-}$.
- The utilization of the membrane cell process is imperative in this context due to the potential harm that active chlorine may cause for the ion-exchange resins in the secondary brine purification process unit (Ali et al., 2018; Lakshmanan & Murugesan, 2014).

To achieve the desired outcome, it is customary to lower the pH of the brine to a range of 2–2.5 and subsequently subject it to either an air-blown wrapped column or a system with a vacuum operating at 50–60 kPa. This process removes most water-soluble chlorine, resulting in a residual concentration ranging from 10–30 mg/l. The vapors that include chlorine are subsequently recirculated to the unprocessed chlorine collection unit or directed toward the chlorine absorbing unit (Sun et al., 2019). The water evaporating from the dechlorinated salt water is condensed within a cooling apparatus. The condensate can be returned to the brine method or undergo chemical dichlorination.

The membrane cell process achieves full dechlorination through various methods, including the passage of brine via an active carbon bed, catalytic decrease, or chemical reducers such as sulfite. The reported residual levels were less than 0.5 mg/l or close to the permissible level of detection and less than 0.1 mg/l.

The diaphragm system does not necessitate dechlorination treatment as any chlorine that permeates employing the diaphragm reacts with caustic soda present in the catholyte space, forming hypochlorite and chlorate (Costilow et al., 1984).

If saturation is conducted using impure salt, a primary purification phase of the entire brine flow is subsequently implemented. Following this, the pH for the dechlorinated salt water is elevated to an alkaline level using caustic soda to mitigate the solubilization of impurities originating from the salt. Without an alkalization step-by-step before resaturation, and with pure salt for saturation followed by primary cleansing on a small portion of the flow, the purification phase is the only stage where alkalization is implemented.

In Section 3.5.2, brine resaturation utilizing solid salt is covered. Evaporation is used to achieve brine resaturation using solution-mined brine in mercury or membrane cell plants. Sodium sulfate precipitates during this process and can be collected, refined, and utilized in various ways (Suslow, 2000).

The catholyte fluid (10–12 weight percent NaOH, 15–17 weight percent NaCl) used in diaphragm cells is utilized immediately or transported through caustic evaporator units where solid salt & 50 weight percent caustic are recovered combined. Before fresh brine enters the diaphragm electrolyzers, it could have been saturated using recycled solid brine collected by caustic evaporators.

3.5.5. Chlorate Destruction: Membrane Cell Technique

Some membrane cells plants run a chlorate devastation unit before any chlorine removal to lessen the accumulation of chlorate within the brine circuits, which might have detrimental effects regarding the ion-exchanging resins, caustic soda quality, and emission levels to the environment (Saputro et al., 2021; Deborde & Von Gunten, 2008). Methods include the catalytic breakdown of sodium chlorate to chlorine using hydrogen and a decrease of sodium chlorate to chlorine using hydrochloric acid during temperatures greater than 85°C.

3.6. CHLORINE PROCESSING, STORAGE, AND HANDLING

3.6.1. General Description

Before chlorine may be utilized, it typically passes through several cooling, cleansing, drying, compression, and liquefaction operations. It can be used as a dry gas in some situations without liquefaction (Ghanbari et al., 2021). Rarely can it be utilized straight from the electrolyzers. Figure 3.12 shows the typical movement of chlorine through electrolyzers to storage. Typically, the chlorine process transforms hot, moist cell gas into a cold, dry gas. The chlorine gas escaping the electrolyzers becomes saturated by water vapor and is around 80–90 C in temperature. Additionally, it has brine mist, residues of chlorinated hydrocarbons, and pollutants, including N_2, H_2, O_2, and CO_2. With just a few bars separating the pressure surrounding the anolyte from the catholyte, electrolyzers function virtually at the atmosphere's pressure (Fomanyuk et al., 2017).

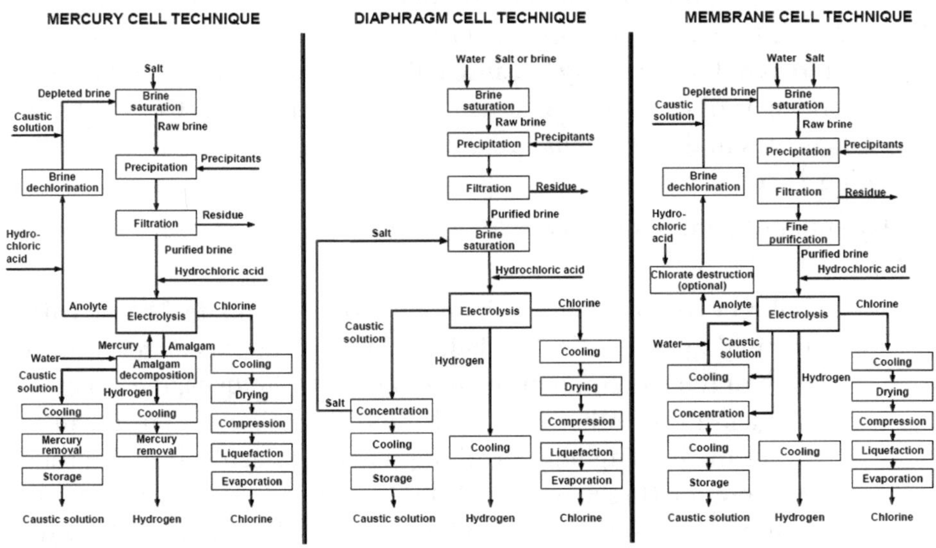

Figure 3.12. *Flow diagram of the three cell techniques.*

Source: Ullmann, Creative Commons License.

3.6.2. Materials

At every stage of processing, based on the working parameters (temperature, pressure, state of the matter, and moisture content), particular consideration must be given to the building materials due to chlorine's strong oxidizing properties. At temperatures under 100°C, most metals are resistant to dry chlorine. Spontaneous ignition occurs at a specified temperature for every metal and also depends on the metal's particle size (150–250°C for iron). The most typical material for dry chlorine gases (water content less than 20 ppm) is carbon steel (Demutskaya & Kalinichenko, 2010). With the possible exception of tantalum & titanium, the material of preference in chlor-alkali plants, wet chlorine gas quickly corrodes the most popular metallic materials. However, titanium spontaneously ignites if the mechanism is not moist enough (ignition temperature 20°C). Other building materials such as metals, graphite, glass, porcelain, and polymers may be employed depending on the circumstances. Unless completely halogenated, oils, and greases often react with chlorine when they come into contact.

3.6.3. Cooling

The overall volume of gas must be handled decreases during the main cooling process, and a significant amount of humidity is condensed (Baldwin, 1927). Water, brine, or other fluids cool things in one or more steps. It is important to avoid overcooling because chlorine and water can mix at about 10 C to create a solid substance known as chlorine hydrate (Cl_2 n H_2O; n = 7–8). Blockages in the procedure are avoided by maintaining a temperature exceeding 15°C.

Two methods are frequently used to cool chlorine gas (Koester & Hites, 1992):

- Indirect cooling using a titanium surface is one technique (often in single-pass vertically shell-and-tube heat exchangers). In the case of the mercurial or membrane cell techniques, the generated condensate is either sent back onto the brine system or, in the event of the diaphragm cell approach, is dechlorinated through evaporation (Preece et al., 2021). This process results in less water that needs to be disposed of since less chlorine is condensed or absorbed. Systems that operate once-through, open-recirculating, and closed-loop can all be used for indirect cooling.
- Another technique involves using water (or the brine or additional fluids) to direct the cooling process. Cooling chlorine gas involves its direct introduction into the lower section of a tower. The water is discharged from the upper section and moves opposite to the chlorine. Typically, the cooling water is devoid of residual

ammonium salts to prevent the occurrence of nitrogen trichloride creation (Morino et al., 2011; Gibson et al., 2008). This approach offers improved mass transfer properties and increases thermal efficiencies. Closed-loop systems are typically utilized for direct cooling.

3.6.4. Cleaning of Wet Chlorine

After the initial cooling process, water droplets and impurities, including brine mist, are eliminated through mechanical means utilizing specialized filters containing glass wool fillings and porous quartz granules or by employing a special electrical precipitator. Subsequently, chlorine is conveyed to the drying towers (Takemura et al., 1994).

3.6.5. Drying

The chlorine present in the cooling system exhibits a state of partial or complete saturation with water vapor. The volumetric water content generally ranges from 1% to 3%. It is imperative to decrease this to prevent subsequent corrosion and to minimize the creation of hydrates.

The process of dehydrating chlorine is predominantly executed through the utilization of concentrated sulfuric acid (96–98 wt.%) in contact towers that operate in a countercurrent manner, spanning two to six stages. This method effectively decreases the moisture level to below 20 mg/m^3. The final moisture content is contingent upon the temperature and concentration in the order of the sulfuric acid utilized during the ultimate drying phase (Donelson et al., 2000). To achieve low-temperature liquefaction (as discussed in Section 3.6.8), it is necessary to reduce the moisture content. This can be accomplished by increasing the number of equilibrium stages in the drying towers or utilizing molecular sieves at 3–9 mg/m^3 levels.

Typically, the quantity of stages is augmented to reduce the ultimate potency of the depleted sulfuric acid. Three stages are required to achieve a used acid concentration ranging from 50–65 wt.%, whereas a 30–40 wt.% result necessitates six stages. The plastic packing present in the columns is designed to resist the corrosive effects of chlorine and sulfuric acid. This feature enhances the distribution of fluids, leading to improved efficiency and reduced pressure drops. Consequently, energy consumption is lowered. The exothermic reaction that occurs during the dilution process of the acid-circulating system is managed through titanium heat exchangers (Maka & Norval, 2013). The acid utilized is then subjected to dechlorination through chemical means or stripping. The spent acid's concentration is contingent upon the number of drying stages alongside the subsequent potential application or mode of elimination. Under certain circumstances, the acid undergoes a reconcentration process to attain a weight percentage of 96% through a vacuum heating method, followed by recirculation. Occasionally, the acid is vented or employed for alternative objectives. On occasion, it may transform into refuse.

3.6.6. Cleaning of Dry Chlorine

Upon exiting the apex of the air-drying tower, anhydrous chlorine undergoes a process wherein it traverses highly effective demisters and a packed bed to impede the entrainment of sulfuric acid droplets (Al-DMooR & El-Qudah, 2016).

Additional cleaning measures may be taken following the process of chlorine drying comprising:

- The process of adsorption upon carbon beds is employed to eliminate organic impurities.
- The technique of absorption-desorption, which involves the use of a suitable solvent like carbon tetrachloride, is utilized to eliminate nitrogen trichloride along with organic impurities.
- Using strong hydrochloric acid to scrub;
- Using liquid chlorine to clean;
- Using ultraviolet light to destroy the nitrogen trichloride and hydrogen.

3.6.7. Compression

According to the rate of production and the required pressure, chlorine gas can be crushed by several compressors following drying and perhaps additional cleaning (Dobry & Abdoun, 2017).

Rotating compressors, like:

- Liquid ring air conditioners for sulfuric acid can handle processing speeds of 150 t/d for each compressor and operate at pressures of 4 bar. Alternatively, two-stage compressors can be utilized to achieve pressures of 12 bar (Dobry & Abdoun, 2015;).
- Screw compressors are suitable for low processing speeds and can operate at forces of upward of 16 bars.

- Reciprocating compressors, a type of mechanical device used to compress gases, such as air, are commonly employed in various industrial applications:
- Dry ring compressors can handle speeds of 200 t/d for each compressor and operate at pressures as high as 16 bars.
- Centrifugal compressors are a type of compressor commonly used in industrial applications:
- Turbo compressors can operate in mono or multi-stage configurations, with a capacity of approximately 1,800 tons per day per unit and a maximum pressure of 16 bar.
- Sundyne blowers can handle throughputs ranging from 80 to 250 tons per day per compressor, with a maximum pressure of 3 bar (Xi et al., 2022).

Multi-stage units with intercoolers are typically required due to the thermal accumulation resulting from compression. Typically, compressor seals are equipped with a pressurized purge system to prevent the escape of chlorine into the surrounding environment. When exposed to high temperatures, iron can undergo spontaneous and uncontrolled reactions with dry chlorine (Liu et al., 2022). Typically, temperatures over chlorine are maintained at levels below 120°C.

3.6.8. Liquefaction

The process of liquefaction can be achieved through various pressure and temperature conditions, including but not limited to ambient temperature as well as elevated pressure (e.g., 18°C and 7–12 bar), low temperatures, and low pressure (e.g., –35°C and 1 bar), or any other possible intermediate arrangement of pressure and temperature (de Carvalho et al., 2015). The selection of appropriate liquefaction conditions is contingent upon several crucial factors, including the chemical makeup of the chlorine gas, the desired level of liquid chlorine simplicity, and the desired yield. Elevating the liquefaction pressure leads to a rise in compression energy consumption while concurrently reducing the energy required for cooling, culminating in a total reduction in energy use.

The maximum liquefaction yield is generally restricted to 90–95% in a one-step process. This is because the residual gas contains a high concentration of hydrogen, which must be maintained beneath the less explosive limit (de Carvalho et al., 2016).

Multi-stage liquefaction has the potential to achieve greater yields of as high as 99.8%. Small-volume liquefiers designed to prevent explosions are commonly utilized following the primary liquefaction process. An inert gas is also introduced to maintain the combination below the less explosive threshold. An alternative approach involves the elimination of hydrogen coming from the system through a reaction with chlorine dioxide within a column. This process produces hydrogen chloride, which can be retrieved from a hydrochloric acid extraction unit. Subsequently, the residual chlorine gas can be condensed securely (Brown & Wardowski, 1984). This approach may be deemed suitable in cases where hydrochloric acid holds commercial value or serves as a raw material for subsequent manufacturing processes, such as the production of ferric chloride.

The selection of the refrigerant for a specific phase of the liquefaction process is contingent upon the pressure for the chlorine. Water has the potential to function as a passive refrigerant under conditions of significant pressure.

In situations where the pressure is comparatively low, alternative refrigerants that include hydrochlorofluorocarbons (HCFCs), specifically chlorodifluoromethane (HCFC-22) and 1,1,1,2-tetrafluoromethane (HFC-134a) (for indirect cooling), ammonia (for indirect cooling), or liquid chlorine (for direct cooling) are employed (Hammer & Gregory, 1964).

The utilization of HCFCs, specifically HCFC-22, is typically proscribed. However, HCFCs that have been reclaimed or recycled may be employed to maintain or service extant refrigeration apparatus until the 31st of December in the year 2014.

Two surveys were conducted in the years 2010 and 2012, wherein 24 plants located in the EU-27 region furnished data about the refrigerants employed for liquefaction during the period spanning from 2008 to 2011. Among the plants surveyed, eight of them utilized HCFC-22, while seven of them employed HFC-134A. Additionally, five of the plants utilized R-507A, three of them utilized ammonia, and two of them utilized carbon dioxide. Each refrigerant, namely chlorine, water, R-410A, and R-422A, was utilized by a single plant in each instance. Several botanical specimens employ a blend of refrigerants to effectuate the liquefaction of chlorine (Giauque & Powell, 1939). The thermal state of chlorine gas during a specific phase is predominantly influenced by the pressure level after compression. Water cooling is typically facilitated by a pressure level exceeding 8 bar. However, it is important to note that such a pressure level also entails a heightened level of risk. Table 3.3 displays the potential trade-offs that may arise from utilizing different kinds of chlorine gas melting systems and refrigeration fluids.

Table 3.3. Trade-Offs in Chlorine Gas Liquefaction

Liquefaction System	Refrig-erant	Safety Aspect	Costs ([1])
The experimental conditions involved elevated pressure levels ranging from 7 to 16 bar and elevated temperatures of approximately 40°C (Saroha, 2006).	Water	High precautions	Low energy costs
The operating conditions involve moderate pressure levels ranging from 2 to 6 bar and moderate temperatures within the range of –10°C to –20°C.)	Water, HCFC/HFC or ammonia.	Moderate precautions.	Moderate energy costs.
Conditions of ambient pressure, approximately 1 bar, and subzero temperatures, specifically below –40°C.	Mainly HCFC/HFC or ammonia	Precautions ([2])	High energy costs

Source: Serge Roudier, Creative Commons License.

The residual chlorine in the tail gas can produce hypochlorite, iron (III) chloride, or hydrochloric acid. The residual chlorine, which cannot be valorized, is then led to the chlorine absorption unit (Shah et al., 1999). An absorption-desorption process with carbon tetrachloride (CCl_4) sometimes recovers it. The latter has the disadvantage of using a toxic substance with a high ozone depletion and global warming potential (GWP).

3.6.9. Handling and Storage

The storage of liquefied chlorine typically occurs at either ambient or low temperatures. The pressure observed is directly related to the vapor pressure exhibited by the compressed chlorine at the specific temperature prevailing within the storage tank (Gaines & Donelson, 1982). The utilization of pressure preservation at ambient temperatures, which typically involves a pressure of approximately 7 bar at a temperature of 20°C, offers several benefits, such as operational simplicity, convenient external visual inspections, and reduced energy and investment expenses. The operation of low-pressure preservation at or near the boiling temperature of liquid chlorine, which is –34°C, necessitates a more intricate infrastructure, specific safety protocols, and increased energy expenditures.

Chlorine can be conveyed through pipelines, either in gaseous or liquid form, over considerable distances within a plant spanning several kilometers (Zhang et al., 2020). The liquid chlorine sourced from the bulk tank has the potential to serve as a feedstock over on-site processes or may be conveniently loaded into containers, road, or rail tankers for transportation purposes.

3.6.10. Vaporization

The vaporization process is typically a preliminary step in utilizing liquid chlorine. One feasible approach involves utilizing ambient heat to vaporize roughly 5 kilograms of chloride per hour and square foot of container surface. To achieve elevated flow rates, it is imperative to utilize a chlorine vaporizer (Chemla et al., 2004).

3.7. SUMMARY

The chlor-alkali sector mainly manufactures chlorine, caustic soda, and hydrogen using an electrolytic technique called chlor-alkali. The industry above holds significant importance in diverse sectors, including chemical compounds, paper and pulp production, fabric, detergents, and soaps, treatment of water, and other related areas. The industry frequently employs various processes and techniques, including the mercury cell process, diaphragmatic cell procedure, membrane cell process, brine purification, and product storage and handling. The chlor-alkali sector has experienced substantial progressions and transformations, particularly transitioning from mercury compound cell technology to more ecologically sustainable methods.

REFERENCES

1. Abd-Almohi, H. H., Alismaeel, Z. T., & M-Ridha, M. J., (2022). Broad-ranging review: Configurations, membrane types, governing equations, and influencing factors on microbial desalination cell technology. *Journal of Chemical Technology and Biotechnology, 97*(12), 3241–3270.
2. Ahmad, N., & Baddour, R. E., (2014). A review of sources, effects, disposal methods, and regulations of brine into marine environments. *Ocean and Coastal Management, 87*, 1–7.
3. Al-DMooR, H. M., & El-Qudah, J. M., (2016). Cake flour chlorination and alternative treatments. *Curr. Res. Nutr. Food Sci., 4*, 127–134.
4. Ali, J. M., Hussain, M. A., Tade, M. O., & Zhang, J., (2015). Artificial intelligence techniques applied as estimator in chemical process systems – A literature survey. *Expert Systems with Applications, 42*(14), 5915–5931.
5. Ali, S. N., Arif, H., Khan, A. A., & Mahmood, R., (2018). Acute renal toxicity of sodium chlorate: Redox imbalance, enhanced DNA damage, metabolic alterations and inhibition of brush border membrane enzymes in rats. *Environmental Toxicology, 33*(11), 1182–1194.
6. Altheide, T., Chevrier, V., Nicholson, C., & Denson, J., (2009). Experimental investigation of the stability and evaporation of sulfate and chloride brines on Mars. *Earth and Planetary Science Letters, 282*(1–4), 69–78.
7. Baldwin, R. T., (1927). Uses of chlorine. *Journal of Chemical Education, 4*(4), 454.
8. Banerjee, S., & Curtin, D. E., (2004). Nafion® perfluorinated membranes in fuel cells. *Journal of Fluorine Chemistry, 125*(8), 1211–1216.
9. Barros, A. I., Gonçalves, A. L., Simões, M., & Pires, J. C., (2015). Harvesting techniques applied to microalgae: a review. *Renewable and Sustainable Energy Reviews, 41*, 1489–1500.
10. Belhadj Ammar, R., Ounissi, T., Baklouti, L., Larchet, C., Dammak, L., Mofakhami, A., & Selmane Belhadj, H. E., (2022). A new method based on a zero gap electrolysis cell for producing bleach: Concept validation. *Membranes, 12*(6), 602.
11. Chemla, M., Durand-Vidal, S., Zanna, S., Petitdidier, S., & Levy, D., (2004). Silicon surface wet cleaning and chemical oxide growth by a novel treatment in aqueous chlorine solutions. *Electrochimica Acta, 49*(21), 3545–3553.
12. Costilow, R. N., Uebersax, M. A., & Ward, P. J., (1984). Use of chlorine dioxide for controlling microorganisms during the handling and storage of

fresh cucumbers. *Journal of Food Science, 49*(2), 396–401.

13. Crook, J., & Mousavi, A., (2016). The chlor-alkali process: A review of history and pollution. *Environmental Forensics, 17*(3), 211–217.
14. Cussler Jr, E. L., & Dunlop, P. J., (1966). An experimental comparison of the Gouy and the diaphragm cell methods for studying isothermal ternary diffusion. *The Journal of Physical Chemistry, 70*(6), 1880–1888.
15. Dalvi, A. G. I., Mohammad, N. K., Al-Sulami, S., Sahul, K., & Al-Rasheed, R., (2000). Role of chemical constituents in recycle brine on the performance of scale control additives in MSF plants. *Desalination, 129*(2), 173–186.
16. de Carvalho, A. L., Antunes, C. H., Freire, F., & Henriques, C. O., (2015). A hybrid input–output multi-objective model to assess economic–energy–environment trade-offs in Brazil. *Energy, 82*, 769–785.
17. de Carvalho, A. L., Antunes, C. H., Freire, F., & Henriques, C. O., (2016). A multi-objective interactive approach to assess economic-energy-environment trade-offs in Brazil. *Renewable and Sustainable Energy Reviews, 54*, 1429–1442.
18. De Nora, V., (1950). The voltage balance in the mercury cell. *Journal of The Electrochemical Society, 97*(10), 346.
19. De Vito, C., Mignardi, S., Ferrini, V., & Martin, R. F., (2011). Reject brines from desalination as possible sources for environmental technologies. In: *Expanding Issues in Desalination* (pp. 85–102). InTech.
20. Deborde, M., & Von Gunten, U. R. S., (2008). Reactions of chlorine with inorganic and organic compounds during water treatment-kinetics and mechanisms: A critical review. *Water Research, 42*(1–2), 13–51.
21. Demutskaya, L. N., & Kalinichenko, I. E., (2010). Photometric determination of ammonium nitrogen with the Nessler reagent in drinking water after its chlorination. *Journal of Water Chemistry and Technology, 32*, 90–94.
22. Dobry, R., & Abdoun, T., (2015). Cyclic shear strain needed for liquefaction triggering and assessment of overburden pressure factor k σ. *Journal of Geotechnical and Geoenvironmental Engineering, 141*(11), 04015047.
23. Dobry, R., & Abdoun, T., (2017). Recent findings on liquefaction triggering in clean and silty sands during earthquakes. *Journal of Geotechnical and Geoenvironmental Engineering, 143*(10), 04017077.
24. Donelson, J. R., Gaines, C. S., & Finney, P. L., (2000). Baking formula innovation to eliminate chlorine treatment of cake flour. *Cereal Chemistry, 77*(1), 53–57.

25. dos Santos Garcia, C., Meincheim, A., Junior, E. R. F., Dallagassa, M. R., Sato, D. M. V., Carvalho, D. R., & Scalabrin, E. E., (2019). Process mining techniques and applications – A systematic mapping study. *Expert Systems with Applications, 133*, 260–295.
26. Dötzel, O., & Schneider, L., (2002). Non-asbestos diaphragms in chlor-alkali electrolysis. *Chemical Engineering and Technology, 25*(2), 167–171.
27. Easteal, A. J., Price, W. E., & Woolf, L. A., (1989). Diaphragm cell for high-temperature diffusion measurements. Tracer diffusion coefficients for water to 363 K. *Journal of the Chemical Society, Faraday Transactions 1: Physical Chemistry in Condensed Phases, 85*(5), 1091–1097.
28. Erxleben, H., & Ruzicka, J., (2005). Atomic absorption spectroscopy for mercury, automated by sequential injection and miniaturized in lab-on-valve system. *Analytical Chemistry, 77*(16), 5124–5128.
29. Fomanyuk, S. S., Kolbasov, G. Y., Chernii, V. Y., & Tretyakova, I. N., (2017). Gasochromic α, β-Ni $(OH)_2$ films for the determination of CO and chlorine content. *Sensors and Actuators B: Chemical, 244*, 717–726.
30. Gaines, C. S., & Donelson, J. R., (1982). Contribution of chlorinated flour fractions to cake crumb stickiness. *Cereal Chem, 59*(5), 378–380.
31. Garcia-Herrero, I., Margallo, M., Onandía, R., Aldaco, R., & Irabien, A., (2017). Life cycle assessment model for the chlor-alkali process: A comprehensive review of resources and available technologies. *Sustainable Production and Consumption, 12*, 44–58.
32. Gardiner, W. C., (1953). Electrolytic caustic and chlorine industries. *Journal of Chemical Education, 30*(3), 116.
33. Gardiner, W. C., (1978). Advances in electrolytic production of industrial chemicals from 1952 to 1977. *Journal of the Electrochemical Society, 125*(1), 22C.
34. Ghanbari, S., Fatehizadeh, A., Khiadani, M., Taheri, E., & Iqbal, H. M., (2022). Treatment of synthetic dye containing textile raw wastewater effluent using UV/chlorine/Br photolysis process followed by activated carbon adsorption. *Environmental Science and Pollution Research, 29*(26), 39400–39409.
35. Giauque, W. F., & Powell, T. M., (1939). Chlorine. The heat capacity, vapor pressure, heats of fusion and vaporization, and entropy. *Journal of the American Chemical Society, 61*(8), 1970–1974.
36. Gibičar, D., Horvat, M., Logar, M., Fajon, V., Falnoga, I., Ferrara, R., & Pacyna, J., (2009). Human exposure to mercury in the vicinity of chlor-alkali plant. *Environmental Research, 109*(4), 355–367.

37. Govindarajan, V. S., & Connell, D. W., (1983). Ginger – Chemistry, technology, and quality evaluation: Part 1. *Critical Reviews in Food Science and Nutrition, 17*(1), 1–96.
38. Gude, V. G., (2015). Energy storage for desalination processes powered by renewable energy and waste heat sources. *Applied Energy, 137*, 877–898.
39. Hammer, R. R., & Gregory, N. W., (1964). Vaporization reactions in the copper chloride-chlorine system. *The Journal of Physical Chemistry, 68*(11), 3229–3233.
40. Hanson, A., Zachritz, W., Stevens, K., Mimbela, L., Polka, R., & Cisneros, L., (2004). Distillate water quality of a single-basin solar still: Laboratory and field studies. *Solar energy, 76*(5), 635–645.
41. Hocking, M. B., & Hocking, M. B., (1985). Electrolytic sodium hydroxide and chlorine and related commodities. *Modern Chemical Technology and Emission Control*, 134–154.
42. Huang, X., Qu, Y., Cid, C. A., Finke, C., Hoffmann, M. R., Lim, K., & Jiang, S. C., (2016). Electrochemical disinfection of toilet wastewater using wastewater electrolysis cell. *Water Research, 92*, 164–172.
43. Ilyas, S., Srivastava, R. R., & Kim, H., (2021). O_2-enriched microbial activity with pH-sensitive solvo-chemical and electro-chlorination strategy to reclaim critical metals from the hazardous waste printed circuit boards. *Journal of Hazardous Materials, 416*, 125769.
44. Jakšić, M. M., Jovanović, D. R., & Csonka, I. M., (1968). Depolarization of sodium amalgam decomposition with cobaltized graphite in the mercury-cell process. *Electrochimica Acta, 13*(11), 2077–2087.
45. Jeanmonod, G., & Diethelm, S., (2021). Poisoning effects of chlorine on a solid oxide cell operated in co-electrolysis. *Journal of Power Sources, 506*, 230247.
46. Kaplan, R., Mamrosh, D., Salih, H. H., & Dastgheib, S. A., (2017). Assessment of desalination technologies for treatment of a highly saline brine from a potential CO_2 storage site. *Desalination, 404*, 87–101.
47. Khan, M. T., Hong, P. Y., Nada, N., & Croue, J. P., (2015). Does chlorination of seawater reverse osmosis membranes control biofouling? *Water Research, 78*, 84–97.
48. Kinsey, J. S., Anscombe, F. R., Lindberg, S. E., & Southworth, G. R., (2004). Characterization of the fugitive mercury emissions at a chlor-alkali plant: Overall study design. *Atmospheric Environment, 38*(4), 633–641.
49. Koester, C. J., & Hites, R. A., (1992). Wet and dry deposition of chlorinated dioxins and furans. *Environmental Science and Technology, 26*(7), 1375–

1382.

50. Krol, J. J., Wessling, M., & Strathmann, H., (1999). Concentration polarization with monopolar ion exchange membranes: Current–voltage curves and water dissociation. *Journal of Membrane Science, 162*(1–2), 145–154.

51. Lakshmanan, S., & Murugesan, T., (2014). The chlor-alkali process: Work in progress. *Clean Technologies and Environmental Policy, 16*, 225–234.

52. Landis, M. S., Keeler, G. J., Al-Wali, K. I., & Stevens, R. K., (2004). Divalent inorganic reactive gaseous mercury emissions from a mercury cell chlor-alkali plant and its impact on near-field atmospheric dry deposition. *Atmospheric Environment, 38*(4), 613–622.

53. Langdon, C. M., Bagshaw, J. C., & MacRae, T. H., (1990). Tubulin isoforms in the brine shrimp, Artemia: primary gene products and their posttranslational modification. *European Journal of Cell Biology, 52*(1), 17–26.

54. LeRoy, R. L., (1983). Industrial water electrolysis: present and future. *International Journal of Hydrogen Energy, 8*(6), 401–417.

55. Li, Y. C., Yan, Z., Hitt, J., Wycisk, R., Pintauro, P. N., & Mallouk, T. E., (2018). Bipolar membranes inhibit product crossover in CO_2 electrolysis cells. *Advanced Sustainable Systems, 2*(4), 1700187.

56. Li, Y. C., Zhou, D., Yan, Z., Gonçalves, R. H., Salvatore, D. A., Berlinguette, C. P., & Mallouk, T. E., (2016). Electrolysis of CO_2 to syngas in bipolar membrane-based electrochemical cells. *ACS Energy Letters, 1*(6), 1149–1153.

57. Littel, R. J., Versteeg, G. F., & Van Swaaij, W. P., (1992). Diffusivity measurements in some organic solvents by a gas-liquid diaphragm cell. *Journal of Chemical and Engineering Data, 37*(1), 42–45.

58. Luo, X., Li, X., Wei, C., Shi, X., Zheng, S., & Deng, Z., (2021). Use of carbon dioxide to enhance the brine purification and flocculation performance of PAM flocculants. *Separation and Purification Technology, 267*, 118676.

59. Lyu, X., Tao, R., & Zhang, T., (2024). Pre-combustion mercury removal with co-production of hydrogen via coal electrolysis. *Journal of Environmental Sciences, 136*, 382–389.

60. Maka, K., & Norval, G. W., (2013). Optimizing chlorine compression design with vapor phase nitrogen trichloride destruction. *The Canadian Journal of Chemical Engineering, 91*(4), 718–724.

61. Morino, H., Fukuda, T., Miura, T., & Shibata, T., (2011). Effect of low-

concentration chlorine dioxide gas against bacteria and viruses on a glass surface in wet environments. *Letters in Applied Microbiology, 53*(6), 628–634.

62. Nawrat, G., Buczek, T., Dukowicz, A., Gonet, M., & Kopyto, D., (2003). Non-asbestos diaphragm-cathode systems. *Polish Journal of Chemical Technology, 5*(3), 69–71.
63. Oehmen, A., Vergel, D., Fradinho, J., Reis, M. A., Crespo, J. G., & Velizarov, S., (2014). Mercury removal from water streams through the ion exchange membrane bioreactor concept. *Journal of Hazardous Materials, 264*, 65–70.
64. Oehmen, A., Viegas, R., Velizarov, S., Reis, M. A., & Crespo, J. G., (2006). Removal of heavy metals from drinking water supplies through the ion exchange membrane bioreactor. *Desalination, 199*(1–3), 405–407.
65. Preece, D., Hong Ng, T., Tong, H. K., Lewis, R., & Carré, M. J., (2021). The effects of chlorination, thickness, and moisture on glove donning efficiency. *Ergonomics, 64*(9), 1205–1216.
66. Rath, K. M., & Rousk, J., (2015). Salt effects on the soil microbial decomposer community and their role in organic carbon cycling: A review. *Soil Biology and Biochemistry, 81*, 108–123.
67. Sanders, H. J., Gardiner, W. C., & Wood, J. L., (1953). staff-industry collaborative report mercury cell chlorine and caustic. *Industrial and Engineering Chemistry, 45*(9), 1824–1835.
68. Saputro, S., Yoshimura, K., Takehara, K., Matsuoka, S., & Narsito., (2011). Oxidation of chromium(III) by free chlorine in tap water during the chlorination process studied by an improved solid-phase spectrometry. *Analytical Sciences, 27*(6), 649–652.
69. Saroha, A. K., (2006). Safe handling of chlorine. *Journal of Chemical Health and Safety, 13*(2), 5–11.
70. Schneiders, K., & Luke, B., (1992). Safety and economy of membrane cell electrolyzers. *Modern Chlor-Alkali Technology, 5*, 105–116.
71. Shah, J., & Qureshi, N., (2008). Chlorine gas vs. sodium hypochlorite: What is the best option? *Opflow, 34*(7), 24–27.
72. Shah, N., Rockwell, J., & Huffman, G. P., (1999). Conversion of waste plastic to oil: direct liquefaction versus pyrolysis and hydroprocessing. *Energy and Fuels, 13*(4), 832–838.
73. Shen, Y. M., & Wang, T. G., (2011). NaCl brine preparation from distiller waste and Na_2SO_4. In: *Advanced Materials Research* (Vol. 233, pp. 897–902).

74. Snyder, L. D., Miller, N. H., & Stavins, R. N., (2003). The effects of environmental regulation on technology diffusion: The case of chlorine manufacturing. *American Economic Review, 93*(2), 431–435.

75. Southworth, G. R., Lindberg, S. E., Zhang, H., & Anscombe, F. R., (2004). Fugitive mercury emissions from a chlor-alkali factory: Sources and fluxes to the atmosphere. *Atmospheric Environment, 38*(4), 597–611.

76. Stokes, R. H., (1950). An improved diaphragm-cell for diffusion studies, and some tests of the method. *Journal of the American Chemical Society, 72*(2), 763–767.

77. Stokes, R. H., (1951). Integral diffusion coefficients of potassium chloride solutions for calibration of diaphragm cells. *Journal of the American Chemical Society, 73*(7), 3527–3528.

78. Suleman, F., Dincer, I., & Agelin-Chaab, M., (2016). Comparative impact assessment study of various hydrogen production methods in terms of emissions. *International Journal of Hydrogen Energy, 41*(19), 8364–8375.

79. Sun, X., Baldwin, E., & Bai, J., (2019). Applications of gaseous chlorine dioxide on postharvest handling and storage of fruits and vegetables – A review. *Food Control, 95*, 18–26.

80. Suslow, T., (2000). Chlorination in the production and postharvest handling of fresh fruits and vegetables. *Fruit and Vegetable Processing*, 2–15.

81. Takemura, Y., Seno-o, K., Mukai, T., & Suzuki, M., (1994). Decomposing organic chlorine compounds in dry cleaning wastewater by Fenton's reaction on reticulated iron. *Water Science and Technology, 30*(3), 129.

82. van Hemert, F. J., Amons, R., Pluijms, W. J., van Ormondt, H., & Möller, W., (1984). The primary structure of elongation factor EF-1 alpha from the brine shrimp Artemia. *The EMBO Journal, 3*(5), 1109–1113.

83. Vanhaecke, P., Vrieze, L. D., Tackaert, W., & Sorgeloos, P., (1990). The use of decapsulated cysts of the brine shrimp Artemia as direct food for carp *Cyprinus carpio* L. larvae. *Journal of the World Aquaculture Society, 21*(4), 257–262.

84. Vargas-Barbosa, N. M., Geise, G. M., Hickner, M. A., & Mallouk, T. E., (2014). Assessing the utility of bipolar membranes for use in photoelectrochemical water-splitting cells. *ChemSusChem, 7*(11), 3017–3020.

85. Wang, L., & Li, L., (2023). Preparation of PVC-LMZO membrane and its lithium adsorption performance from brine. *Desalination, 561*, 116689.

86. Xi, D., Wen, S., Zhang, X., Xie, W., Fang, Z., Zhou, R., & Ostrikov, K. K., (2022). Plasma-electrolytic liquefaction of human waste for biofuels

production and recovery of ammonium, chlorine, and metals. *Chemical Engineering Journal, 433*, 134581.

87. Xu, J., Amorim, I., Li, Y., Li, J., Yu, Z., Zhang, B., & Liu, L., (2020). Stable overall water splitting in an asymmetric acid/alkaline electrolyzer comprising a bipolar membrane sandwiched by bifunctional cobalt-nickel phosphide nanowire electrodes. *Carbon Energy, 2*(4), 646–655.
88. Zaviska, F., Drogui, P., & Pablo, G., (2012). Statistical optimization of active chlorine production from a synthetic saline effluent by electrolysis. *Desalination, 296*, 16–23.
89. Zhang, B. S., Lv, X. F., Zhang, Z. X., Liu, Y., Kim, J. K., & Xin, Z. X., (2010). Effect of carbon black content on microcellular structure and physical properties of chlorinated polyethylene rubber foams. *Materials and Design, 31*(6), 3106–3110.

CHAPTER

4 Anodes in Chlor-Alkali Industry

Contents

4.1. INTRODUCTION

The chlorine evolution process's anode area circumstances need strict anode material requirements. Stability against electrochemical oxidation and chemical assault by KCl, NaCl, HCl, HOCl, and others is important (Karlsson & Cornell, 2016).

Given the tendency of surface films to dissolve or develop, maintaining surface layer conductivity at potentials that favor chlorine production is a concern. Electrocatalysis reduces anodic overpotential.

Platinum, magnetite, and carbon anodes were used in chlorine electrolysis. Platinum's high cost and magnetite anodes' poor machinability, fracture, and conductivity were noted. Magnetite has 5% of the electrical conductivity of graphite. Magnetite's anode barely reached 0.4 kAm^{-2}. The synthetic graphite production technique, invented by Castner, Chas, and Acheson, increased graphite use from the early 20th century through the late 1960s. Graphite anodes had a 6–24-month lifetime, which was a drawback. The abrasive effects of anode gases and the carbon-to-carbon monoxide oxidation process need 1 to 3 kilos of graphite per ton of chlorine, causing this phenomenon. Product stream chlorinated hydrocarbons indicate contaminants (Pilla et al., 1997). Graphite particles impair diaphragm functioning. Cell electrode machinability and setup are difficult.

As the anode degraded, the distance between it and the cathode increased, raising cell voltage and energy use. Graphite deterioration reduction attempts have failed.

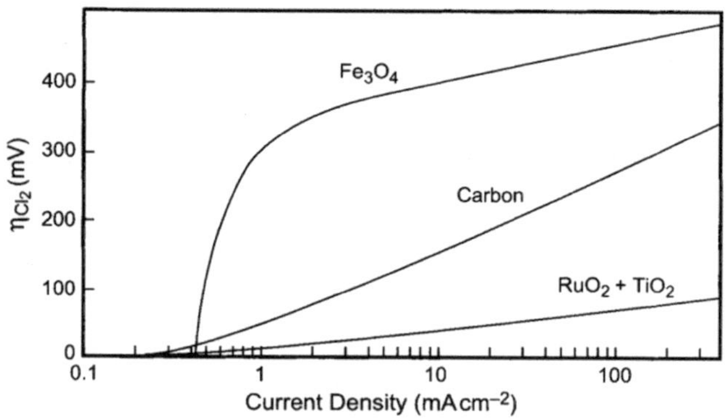

Figure 4.1. *This study investigates chlorine overvoltage and current density on numerous anodes in 5M NaCl solutions at 95°C and pH 3.5.*

Source: J. Billiter, Creative Commons License.

The academic community recognizes platinum group metals' electrical conductivity and stability in the anolyte environment. The formation of strongly adhering, protective oxide coatings that shield metallic substances from corrosion is also documented (Figure 4.1).

The endeavor to integrate these two characteristics into an anode configuration was instigated in 1913 by Stevens via the development of anodes made of tungsten coated with Pt. During the early 1960s, an experimental investigation was conducted to examine the application of titanium anodes coated with platinum. However, the investigation indicated that the approach did not produce desirable results. Platinum's wear rates were discovered to be excessively high, and the electrodes displayed a tendency towards early coating losses, particularly in the context of mercury cells. Further inquiries have led to the identification of coatings composed of RuO_2 and TiO_2 on titanium substrates (O'Brien et al., 2005). The electrodes are known as dimensionally stable anodes (DSAs) due to their capacity to withstand deterioration. The main body of literature regarding coatings originating from Beer's RuO_2 was disseminated through the publication of South African Patents 6,62,667 (1966) and 6,80,034 (1968). Subsequently, patents were issued in the United States encompassing various features of solid solutions comprising $RuO_2 + TiO_2$. The patents in question are U.S. Patents 3214110, 3632498, 3711385, and 3751291. The solid solution coating composed of RuO_2 and TiO_2 was refined, leading to its eventual commercialization. Diamond Shamrock Technologies, S.A., located in Geneva, conducted the worldwide licensing of the product (Lakshmanan & Murugesan, 2014).

4.2. ELECTRODE PREPARATION

The electrode known as DSA is composed of a titanium substrate coated with a combination of RuO_2 and TiO_2. The coating exhibits a precious metal oxide composition of up to 50 mol%, with the quantity of precious metal loading varying between 5 and 20 gm^{-2}. As described in the patent literature, the coatings are typically generated using the thermal breakdown technique, which is outlined in the following manner (Casimir et al., 2016).

Beer's process, as outlined in South African Patent 6,80,834, involves the treatment of a substrate composed of Grade 1 or 2 titanium with hot aqueous oxalic acid, subsequently followed by cleaning with an ultrasonic cleaner in water. After drying, the electrode is subjected to a coating procedure utilizing a solution comprising butyl alcohol with an approximate concentration of

0.5 M hydrochloric acid, ruthenium in $RuCl_3$, and titanium as butyl titanate (Tong et al., 2008). The ratio of Ru to Ti present in the solution aligns with the targeted value for the coating. Following this, the titanium substrate undergoes a coating procedure which entails the deposition of $RuO_2 + TiO_2$, succeeded by being exposed to temperatures within the range of 300°C to 500°C for 1 to 5 minutes in an ambient air atmosphere.

Utilizing a "mixed-crystal" coating composed of RuO_2, TiO_2, and SnO_2 is a prevalent method for preparing electrodes in diaphragm cells. As a result of reduced Ru loading, the material's cost-effectiveness is enhanced, and it generates a lower amount of O_2 by approximately 22–25% compared to titanium coated with $RuO_2 + TiO_2$. The stoichiometric ratio of $TiO_2/SnO_2/RuO_2$ ranges from 1.5 to 2.5, and SnO_2 constitutes 35–50 mol% of the combined quantity of TiO_2/RuO_2. The process of fabricating the electrode involves the application of a solution containing $RuCl_3$, $SnCl_2$, and HCl onto a purified titanium substrate, followed by thermal treatment at 450°C for 7 minutes under atmospheric conditions (Arunkumar & Paul, 2017). The procedure is iterated 10 times, resulting in an ultimate loading of $RuO_2 + SnO_2$ of approximately 17 g m^{-2}.

4.3. OTHER PATENTED ANODE COMPOSITIONS

As stated, the DSA electrode commonly consists of a valve metal substrate coated with a noble metal-based compound. Significant endeavors have been directed toward creating unique anode coatings to avoid infringing on the Beer patents or improve the efficacy and economic feasibility of anode compositions (Olsen & Thonstad, 1999). The exhaustive examination of the diverse compositions recorded in the literature surpasses the limits of this specific chapter. The simplified schematic depiction of the anode components can be stated as Sub/VM + NM + NNM. "Sub" indicates the substrate, while "VM" denotes the valve metal.

Furthermore, "NM" is representative of the noble metal, and "NNM" indicates the non-noble metal. The literature references various ingredient combinations utilized in the various formulations. The elements mentioned above, namely Zr, Ti, Hf, Ta, Nb, W, Bi, C, Al, and Fe_3O_4, may exist independently or in conjunction with one another or as alloys formed by combining one element with another (Horvick, 2006).

The present study investigates the effects of varying loadings of valve metals (Zr, Ti, Hf, Ta, Nb, W, Bi, Al), noble metals (Ir, Pt, Rh, Os, Ru, Pd), and non-noble metals (Ag, Cu, Au, Co, Fe, Ni, Si, Sn, Pb, As, Cr, Sb, Mn, etc.) on the system. The oxides that were examined in this study include spinels (AB_2O_4), pyrochlores (A_2B_2 O(7-y)), perovskites (ABO_3), and delafossites (AB). Element A has the potential to be classified as a noble metal, whereas element B may be categorized as either a valve metal or a non-noble metal (Mai et al., 2005). The aforementioned generic structures may be occasionally fabricated in layers, such as Sub/NM/VM, and may also be coated with various oxides, such as SiO_2 as disclosed in U.S. Patent 3,677,815.

Numerous patents have documented a wide range of anode compositions. However, only patents about beer have been effectively employed. Table 4.2 presents the current status of patents deemed "valid." All primary patents listed in Table 4.1 have expired and are no longer enforceable (Faes et al., 2012).

Table 4.1. The Current Investigation Centers on the Portrayal of the Principal Anode Patents

General Description	U.S. Patent It	Expiration Year
Application method	3864163	1992
Double-component coatings	3711385	1989
Triple-component coatings	3948751 4003817 4070504 4318795	1993 1994 1995 1999

Source: F. Jeitner, Creative Commons License.

4.4. PHYSICAL CHARACTERISTICS AND MORPHOLOGY OF RUO_2 + TIO_2-BASED ELECTRODES

Despite their classification, metastable phases have been observed to form solid solutions across the complete compositional range on Ti supports at a temperature of 400°C (Pouladvand et al., 2019).

Table 4.2. The Depiction of Certain Crystallographic Characteristics of , , Ti, and Ru

Property		
Symmetry group	/mnm	

Crystal habit	Tetragonal	Tetragonal
Crystal structure	Rutile	Rutile
	0.4594 nm	0.4490 nm
	0.2958 nm	0.3195 nm
V	0.062 n	0.062 n
–	Ti	Ru
Ionic radius	0.068 nm	0.065 nm
Electronegativity	1.5	1.8
Valence	4	4

Source: O DeNora, Creative Commons License.

The system consisting of RuO_2, TiO_2, and Cl was investigated using X-ray diffraction and micro-X-ray spectroscopy. The coexistence of the substance is contingent upon the preparation conditions and is observed to occur with pure RuO_2 on the RuO_2 facet and an anatase solid solution on the TiO_2 facet (Reddy & Ramaprabhu, 2007). The current phase may have a structural flaw and a diminished degree of crystalline order.

The coatings demonstrate crystallite dimensions from 1–5 pm within the temperature range of 300–700°C. Furthermore, the BET and electrochemical area measurements suggest that the coatings exhibit significant surface areas, estimated to be around 200 cm^2 per visible square centimeter, subject to the factors that impact the coating's formulation and composition. The present study involved the utilization of X-ray photoelectron spectroscopy to examine the film terminals of RuO_2 and $RuO_2 + TiO_2$ in 4M NaCl solutions (Jiang e al., 2017). The investigations were carried out before and after anodic polarization. The findings indicate the presence of a defect structure consisting of two discrete species, specifically "Cl" – and adsorbed atomic Cl, in these regions.

The observed structure, characterized as "mud-cracked" through scanning electron microscopy analysis, exhibits a surface with microcracks that form distinct islands. This morphology is attributed to the various factors that influence the preparation of the sample.

Based on the findings presented in Figure 4.2, which was generated utilizing the conductivity data sourced from references, it can be inferred that the conductivity of mixtures containing RuO_2 and TiO_2 exhibits variability in response to alterations in the proportion of RuO_2 present. The results indicate an increase in conductivity beyond a concentration of 25 mol% RuO_2, while a concentration of less than 20 mol% RuO_2 leads to a significant decrease in conductivity. The DSA composition that is deemed commercially viable comprises a maximum of 70–80% TiO_2, thus accounting for its marketability (Kim et al., 2010). The elevated conductivity observed in mixtures comprising more than 25% RuO_2 has been accounted for by forming an unending chain of RuO_2 clusters within the Ti_2 matrix. This explanation is provided despite the intricate nature of the hypothesis concerning the non-metal-to-metal transitions linked to the mixtures of insulators and conductors, which remains a subject of ongoing debate. The inadequate conductivity of mixtures with a RuO_2 content below 25% can be attributed to the chain disruption mentioned above (Mishra & Ramaprabhu, 2011).

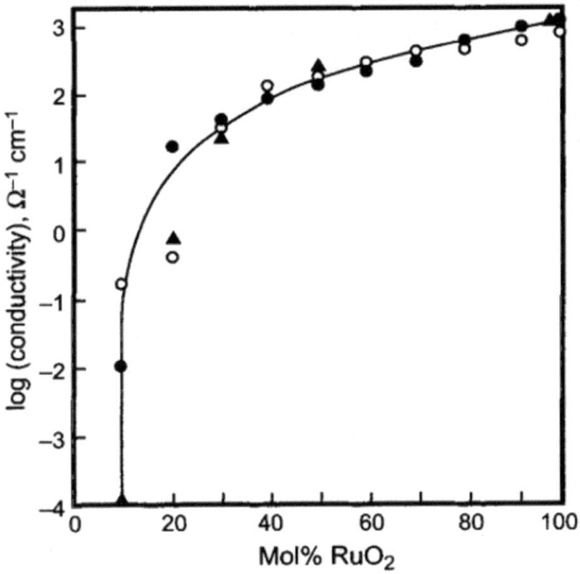

Figure 4.2. *The present study uses direct resistance and impedance measurement techniques to investigate the relationship between the conductivity and mol% in + coatings.*

Source: H. B. Beer, Creative Commons License.

The utilization of spectroscopic techniques in assessing heterogeneous oxide coatings has led to the identification of surface saturation attributed to one of the metals (Sundriyal et al., 2018). The surfaces of (Ti + Ru) O_2 and (Ru + Ir) O_2 exhibit enrichment of Ti and Ir, respectively. The degree of enrichment is contingent upon the method of preparation. Nonetheless, SIMS analysis cannot detect the exterior enrichment of titanium in the RuO_2 + TiO_2 coatings utilized in industry.

4.5. ELECTROCHEMICAL BEHAVIOR OF RUO_2 + TIO_2-BASED COATINGS

Numerous researchers have thoroughly examined the electrochemical characteristics of electrodes made of RuO_2 and have compiled their findings in several papers (Bauerfeind et al., 2022). References, where both the equilibrium and non-equilibrium electrochemical characteristics of oxide electrodes manufactured in diverse ways were investigated, and references and where the mechanical features of the chlorine evolution reactions were rigorously addressed, deserve special notice.

About the reversible hydrogen electrode (RHE) in the identical solution, RuO_2 displays a resting potential of 0.65 V (Feijoo et al., 2023). For a saturated calomel electrode (SCE), the rest potentials of thermally produced RuO_2 films vary linearly with pH having a slope of 0.059 V/pH unit in a pH range of 0–14 (Figure 4.3).

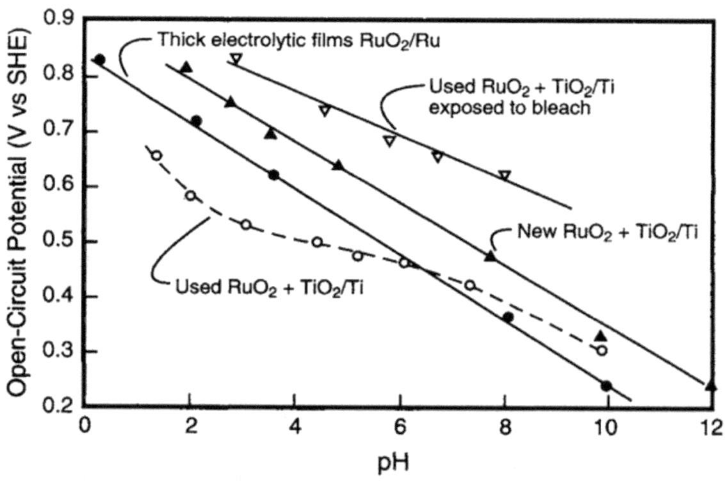

Figure 4.3. *The open circuit potential variation of electrodes with pH at a temperature of 25°C was examined, with the electrodes being thermally formed from titanium and platinum.*

Source: D. M. Novak, Creative Commons License.

This phenomenon has been ascribed to the reaction (Zhu et al., 2018):

$$2RuO_2 + 2H^+ + 2e \leftrightarrow Ru_2O_3 + H_2O$$

Coatings comprising RuO_2 and TiO_2 with a 30 mol% concentration were recently prepared. The RuO_2 material also demonstrates a consistent slope of 60 millivolts per pH. However, electrodes that combine RuO_2 and TiO_2 deviate from this observed trend, as depicted in Figure 4.3 (Burke et al., 1984). Nonetheless, upon exposure of the utilized electrodes to NaOCl solutions, a slope of 60 mV is once again exhibited by the electrodes, indicating the involvement of Cl^- ions in the open circuit behavior:

$$2RuO_2 + 2H^+ + 2e + Cl^- \leftrightarrow Ru_2O_3 \cdot Cl^- + H_2O$$

A 1.2 to 0.4 V voltage range was investigated through cyclic voltametric curves in both acid and alkaline solutions, respectively, as depicted in Figure 4.5. The results indicate that the profile in acidic solutions lacks distinct features, while in alkaline solutions, a well-defined peak is observed before the onset of oxygen evolution (Yao et al., 2001). The electrochemical characteristics of RuO_2 electrodes have been ascribed to a reversible oxidation-reduction process, which operates via a mechanism that entails proton transfer with the solution:

$$RuO_x(OH)_y + dH^+ + de \leftrightarrow RuO_{x-d}(OH)_{y+d}$$

The reaction above is responsible for the notable pseudo-capacitance observed in RuO_2 electrodes, which are utilized to fabricate capacitor devices (Kim & Yang, 2014). Conway's monograph comprehensively analyzes these pseudocapacitors' theoretical underpinnings and real-world implementation (Figure 4.4).

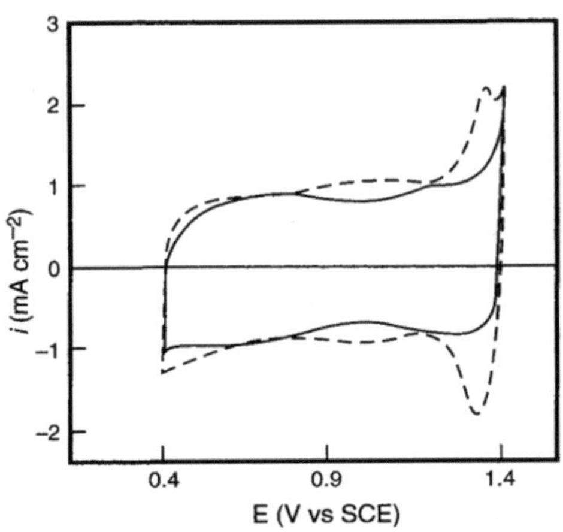

Figure 4.4. *An illustration showing the usual voltametric curves of an electrode in lM solutions of (solid lines) and KOH (dotted lines) with a sweep rate of 20 mV at 25°C.*

Source: D. M. Novak, Creative Commons License.

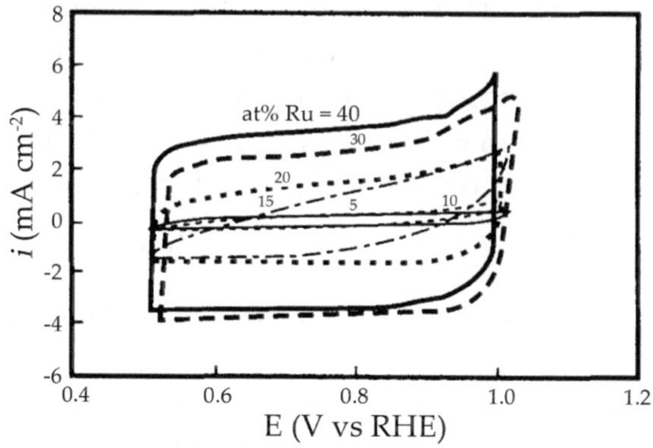

Figure 4.5. *Cyclic voltammetry of new Ru–Ti electrodes of 2 m thickness with varying Ru concentrations in 5M NaCl (pH = 3.5) at 100 mV and 90°C.*

Source: E. N. Balko, Creative Commons License.

The cyclic voltametric curves depicted in Figure 4.6 demonstrate that the current response at a specific potential exhibits fluctuation in proportion to the concentration of Ru in the coating (Zdrachek et al., 2023). The property above can potentially serve as a diagnostic instrument to characterize anodes while they are in operation.

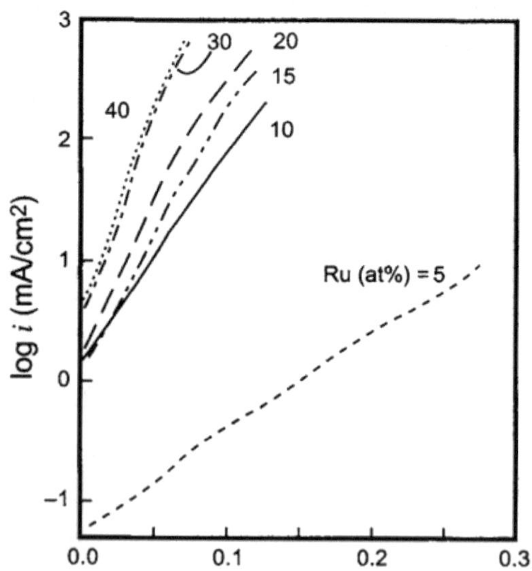

Figure 4.6. *The Tafel plot is affected by the Ru concentration in 5M NaCl with a pH of 3.5 and fresh Ru–Ti electrodes at room temperature.*

Source: Fumio Hine, Creative Commons License.

Figure 4.7 illustrates the polarization characteristics of the RuO_2 + TiO_2 electrode in 5M NaCl solutions during the chlorine evolution process. The electrodes exhibit Tafel slopes of approximately 30 mV when subjected to high Ru levels, whereas electrodes containing less than 10 at% Ru demonstrate a Tafel slope of approximately 120 mV (Betelu et al., 2011). Figure 4.7 displays the exchange current density for the Cl_2 evolution reaction. It has been calculated using the data from Figure 4.6. The results indicate that the value of i_o decreases as the percentage of Ru in the coating decreases.

During the process of chlorine evolution, kinetic studies were conducted using electrodes composed of RuO_2 and TiO_2, with a Ru content of over 25 atomic percent (Tilak et al., 2011). The results of these investigations indicate that the Tafel slope was measured to be 40 millivolts or 2 RT/3F. Additionally, the reaction order concerning chloride ions was determined to be 1, and the stoichiometric number was found to be 1 (Wang et al., 2021). The observations above are not explicable through the conventional pathways as depicted in the equations provided:

$$S + Cl^- \rightarrow S - Cl_{ads} + e$$

$$2S - Cl_{ads} \rightarrow Cl_2 + 2S$$

$$S - Cl_{ads} + Cl^- \rightarrow Cl_2 + e + S$$

The variable "S" utilized in the reactions above pertains to the condition of the surface of the oxide film and not that of the anodic metal (Bommaraju et al., 2008). Various reaction mechanisms have been proposed to explain the experimental observations:

$$S + Cl^- \leftrightarrow S - Cl_{ads} + e$$

$$S - Cl_{ads} \leftrightarrow S - Cl^+{}_{ads} + e$$

$$S - Cl^+{}_{ads} + Cl^- \rightarrow Cl_2 + S$$

$$S + Cl^- \leftrightarrow S - Cl_{ads} + e$$

$$S + H_2O \rightarrow S - OH_{ads} + H^+ + e$$

$$S - Cl_{ads} + S - OH_{ads} \leftrightarrow HOCl + 2S$$

$$HOCl + HCl \leftrightarrow Cl_2 + H_2O$$

Nevertheless, the unexpected discovery of a reaction order of –1 about H^+ prompted the formulation of the reaction mechanism (Wang et al., 2019):

$$S - OH \leftrightarrow S - O + H^+ + e$$

$$S - O + Cl^- \rightarrow O - S + Cl + e$$

$$O - S - Cl + H^+ + Cl^- \rightarrow S - OH + Cl_2$$

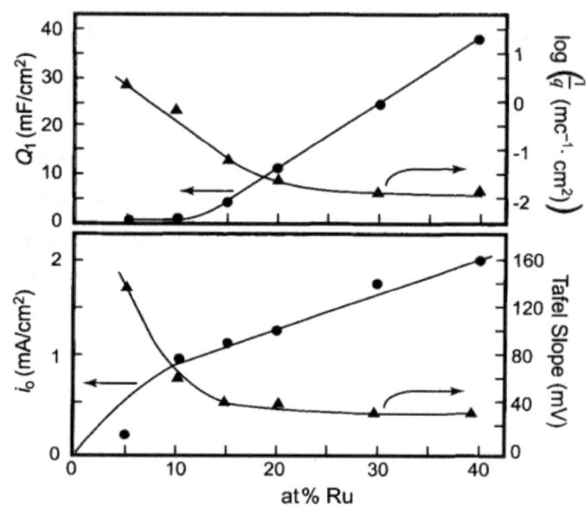

Figure 4.7. *The present study investigates the electrochemical characteristics of Ru–Ti oxide films and their variations. The present study investigates the relationship between capacitance (C) and voltametric charge (q) with the percentage of ruthenium (% Ru). The dependence of %Ru on the bottom and Tafel slope is investigated.*

Source: Thomas F. O'Brien, Creative Commons License.

The scheme above incorporates a reaction order of –1. However, it necessitates a reduction in the pace of the chlorine evolution reaction as the pH decreases, which is inconsistent with empirical observations in the industrial setting (Evdokimov, 2001). The precise understanding of the mechanical facets of the chlorine evolution reaction remains ambiguous.

This study investigates the mechanisms suggested for the chlorine evolution reaction on RuO_2 and RuO_2-based coatings (Tilak et al., 1997). The findings of fundamental kinetic electrochemical investigations indicate that electrodes composed of RuO_2 and TiO_2, with a RuO_2 content exceeding 25%, demonstrate a reduced Tafel slope and an increased exchange current density about the chlorine evolution reaction. The efficacy of said electrodes is contingent upon the preparation methodology, in addition to the concentration of RuO_2 and the loading of Ru metal within the coatings.

The observed phenomenon is that the overvoltage of chlorine experiences an increment as the mole percentage of ruthenium falls below 20, or the loading of ruthenium is below 4 g/m^2 for an electrode containing a 40-mole percentage of ruthenium (Tilak et al., 2001). Consequently, the outcome is a reduced lifespan of the anode.

The temperature at which thermal decomposition occurs plays a crucial role in determining the chlorine level over potential, denoted as ηcl_2. It has been observed that the lowest value of ηcl_2 can be attained by subjecting the substance to baking at a temperature range of 400–500°C for 1 hour. Excessive treatment results in elevated overvoltage that cannot be reversed. It is important to highlight that the elevated values of ηcl_2, which arise from a coating containing less than 20 mol% Ru, are a consequence of the reduced conductivity of the coating (Battisti et al., 2001). The impact of reduced mol% Ru, Ru loading, and inadequate baking conditions are manifested in the operational lifespan. Therefore, to achieve an extended lifespan, it is recommended that the coating contains a ruthenium content exceeding 25 mol% and a Ru loading of over 2 g/m^2.

4.6. CELL PERFORMANCE CHARACTERISTICS WITH RUO_2 + TIO_2-BASED ANODES

The effectiveness of anodes is contingent upon various operational variables, including pH levels, current density, sodium chloride concentration, NaOH (in diaphragmatic and membrane cells), and the quality of brine. As previously noted, the impact of the anode on cell inefficiency is closely linked to the losses that result from the oxygen evolution reaction and the creation of chlorate (Ammar et al., 2023). A decrease in pH occurs concurrently with an elevation in oxygen concentration at the interface of the anode solution. Consequently, a reduction in the overall pH level will lead to a decrease in chlorate production. The oxygen generation at the anode relies on various factors, such as current density, pH level, anode coating composition, surface area, and salt concentration.

The positive correlation between anode efficiency and low anolyte pH values can be attributed to the concomitant chlorate and oxygen concentrations reduction (hi & Misawa, 2021). The efficiency of the diaphragm cell exhibits an increase from approximately 88% to 98% upon reduction of the pH level from 4.8 to 3.35. As per the guidelines stipulated by the DSA manufacturers, utilizing the anodes within a pH spectrum ranging from 2 to 12 is advisable. The phenomenon under observation can be ascribed to the plausible dissolution of the coating in the presence of high pH conditions and the dissolution of titanium in the presence of low pH conditions.

4.6.1. Current Density

The present level of efficacy exhibits an upward trend as the current density rises, as expounded upon (Geuenich et al., 2005). The efficiency of the diaphragm cell is approximately 91% at 1.55 kA/m^2 and 94–95% at 2.32 kA/m^2 when subjected to a caustic strength of 150 g/L NaOH and a temperature of 95°C.

4.6.2. Brine Concentration

The utilization rate of the anode in diaphragm cells experiences a decline as the concentration of NaCl decreases (Nayar et al., 2019). The most favorable feed brine concentration is 315 g/L, while the minimum brine strength is 285 g/L. The reduction in efficiency is observed when the salt concentration falls below 285 g/L, owing to the simultaneous rise in oxygen evolution.

A feed brine concentration of approximately 300 g/L is customary in membrane cell technology. Moreover, it is advised to maintain a minimum salt concentration of 170 g/L in the depleted brine (Afrasiabi & Shahbazali, 2011).

4.6.3. NaOH Concentration

The efficiency of the anodic cell experiences a decline as the strength of NaOH increases in diaphragm cells and with the augmentation of back-migration of OH^{-1} in membrane cells (Koutu et al., 2016).

4.6.4. Temperature

The temperature impacts the voltage of the cell, but it does not affect the efficiency of the cell. The optimal temperature for the anolyte in diaphragm cells is approximately 96°C when a current density of 2.52 kA m^{-2} is applied (Preston-Thomas, 1990).

4.7. ANODE FAILURE MECHANISMS

DSAs for chlorine production employ grade 1 or 2 titanium substrates. The previously mentioned substrates are coated with a composite comprising

RuO_2 and TiO_2, wherein the minimum percentage of RuO_2 present in the mixture is 25%. In specific instances, the integration of SnO_2, RuO_2, and IrO_2 into these coatings is implemented for utilization in the diaphragm and membrane cells (Aurbach et al., 2002). The coatings are commonly produced via the thermal decomposition of mixtures comprising $RuCl_3$, butyl titanate, and butanol at temperatures from 40 to 50°C. The thickness of the coatings is within the range of 10 to 15 micrometers. The roughness factor, the ratio of real area to geometric area, varies from 206 to 300 depending on the thermal decomposition conditions (Mauger, & Julien, 2021). The anodes in diaphragm cells display a variation of 4 to 5 g m^{-2} of Ru, while in membrane and mercury cells, the range is 10–15 g m^{-2}.

In general, the anode potential of newly produced or refurbished anodes, relative to the standard calomel electrode, displays a variation of 1.06 to 1.08 volts when exposed to a current density of 3 to 4 kiloamperes per square meter (Mauger & Julien, 2021). The abovementioned interval indicates a chloride ion overpotential ranging from 95 to 115 millivolts. The point at which an anode becomes electrocatalytically inactive is identified by a rise in anode potential of 200–300 mV. During operation within electrolytic cells, a progressive rise in the anode potential is observed. However, if such an incident occurs in the initial stages, it can be ascribed to insufficient techniques for producing the anode (Wang et al., 2023). The cross-sectional view of an anode coated with an electrocatalytic composition is illustrated in the schematic shown in Figure 4.8.

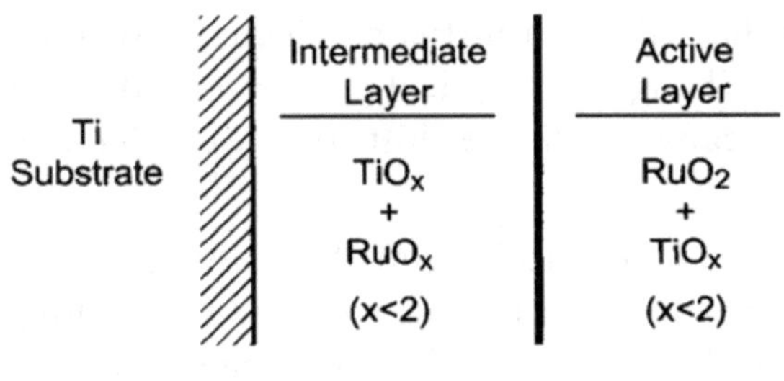

Figure 4.8. *An illustration showing the cross-sectional view of coatings deposited onto titanium.*

Source: J. B. Cotton, Creative Commons License.

Inadequate pre-treatment of the titanium surface before coating may develop titanium surface oxides with an oxygen concentration approaching two. Furthermore, when the potential of the anode is increased, it can lead to the breakdown of titanium oxide films, ultimately resulting in the anodic dissolution of titanium (Kuo et al., 2022). It is essential to guarantee the conductivity of the intermediate layer consisting of Ti and Ru oxides. The process can be facilitated by subjecting the coating to an appropriate thermal treatment. Otherwise, the initial potential of the anode will be increased.

The dependence of pCi_2 and anode lifespan on the mol% of RuO_2 and the quantity of Ru per square meter present in the coating was determined through accelerated testing. This testing was conducted in a 30 gp^1 NaCl solution at a 30-kA m^{-2} current density. An increase in the anode potential and a decrease in its lifespan are observed when the concentration of RuO_2 in the anode falls below 20 mol%. As the molar concentration of RuO_2 approaches 100%, the anode undergoes instability while its potential remains unaffected (Garcia-Sanz et al., 1997). Similarly, the durability of the anode is reduced in Ruthenium coatings containing less than 2 g m^{-2} compared to those with a Ruthenium concentration surpassing 4–5 g m^{-2}. Notably, the magnitude of loading required to attain a prolonged lifespan is contingent upon the proportion of RuO_2 that is present. The inclusion of supplementary dopants, such as SnO_2, and the specific application and firing temperatures employed in these composite oxide coatings are significant variables that warrant consideration.

The increase in anode potential observed due to a reduction in ruthenium concentration while keeping the loading constant can be explained by two fundamental factors (Sen, 2004).

The reduction in the molar percentage of RuO_2 exhibits a direct correlation with the reduction in the conductivity of the composite oxide (Qin et al., 2010). Changes in physical properties and variations in chemical bonding behavior can explain the shift from a conductive metallic state to an insulated state.

The modification of the mechanism of chlorine evolution can be ascribed to a decline in the molar percentage of RuO_2, resulting in a transition from a mechanism based on adsorbed intermediates to a step involving a sluggish transfer of electrons (Maier et al., 2015). The probable cause of this phenomenon can be attributed to the limited accessibility of electrocatalytic sites for the reaction. Notably, when adsorbed intermediates are present, the Tafel slope exhibits a measurement of 30 millivolts. In contrast, in cases where the transfer of charge is impeded, the Tafel slope exhibits a value of 120 millivolts.

The empirical evidence suggests that a reduction in the molar percentage of RuO_2 is associated with an elevation in the Tafel slope, thereby causing a rise in the generation of O_2 (Tan et al., 2017). The presence of a significant % O_2 concentration in real-world situations relies on the appearance of Tafel slopes that surpass 120 mV in every anode connected in parallel in a monopolar cell. When a segment of the anode is rendered inactive owing to inadequate mol% RuO_2 or surface obstruction, the remaining active surface will encounter a substantial current density, leading to a diminished proportion of O_2.

The observed rise in anode potential as RuO_2 loading decreases (while maintaining a constant Ru in the coating) can be attributed to a modification in the mechanism of the chlorine evolution reaction (Carvalho et al., 2003).

The decrease in anode longevity associated with a low initial RuO_2 loading coating can be attributed to the inadequate density of the coating in thin layers (Brightman et al., 2011). This leads to the formation of a TiO_2 layer on the Ti substrate, which contributes to an increase in anode potential.

The rise in chlorine over potential can be attributed to impurities such as MnO_2, SiO_2, $BaSO_4$, Fe_2O_3, etc., contaminating the active surface (Mohtadi et al., 2004). The presence of contaminants can elevate the anode potential by obstructing the active surface or through interaction with RuO_2 and TiO_2, particularly in the case of Fe_2O_3. This can result in the surface enrichment of TiO_2, as evidenced by the following:

$$4FeO + RuO_2 \rightarrow 2Fe_2O_3 + Ru$$

$$Ru + 4H_2O \rightarrow RuO_4^{2-} + 8H^+ + 6e$$

$$FeO + TiO \rightarrow TiO_2 + Fe$$

The phenomenon of surface blockage occurs due to the accumulation of insoluble species or the adsorption of organic substances. Table 4.3 presents observations of impurities that have a detrimental impact on the anodes (Robinson & George, 1954).

In cases where the anode potential is elevated due to blockage, performing a chemical or mechanical cleaning procedure can effectively restore the deactivated anode. If the Tafel slope exceeds 120 mV on anodes entirely coated with impurities, the chlorine evolution reaction may be superseded by O_2 evolution, leading to a high percentage of O_2 (Uribe et al., 2002).

The adverse impacts resulting from the surface sets can be mitigated to a certain degree by employing prudent cleaning techniques utilizing suitable agents such as hydrochloric acid for iron deposits and sodium hydroxide for silicon dioxide deposits (Yokokawa et al., 2011). In instances where the loading diminishes to less than 2 g m^{-2}, or if the percentage of RuO_2 in the coating, particularly in the uppermost layers, falls below 20%, it becomes necessary to reapply the anode coating after removing the previous coating. This is done to enhance the anode potential and overall performance.

4.7.1. Reasons for Lowered Loading RuO_2 and Mol% RuO_2

The gradual removal of active coating layers due to the emission of gases at an anode can result in erosion and subsequent reduction in loading. Simultaneously, the selective dissolution of RuO_2 can occur due to chlorine evolution, outages, and oxygen generation (Monreal et al., 1994). This phenomenon is expected to alter the chemical and physical makeup of the uppermost strata.

Table 4.3. The Impact of Impurities Present in Brine on Anodes

Impurity	Maximum Allowed Concentration (ppm)	The Effect at Greater Concentrations
Ba	0.4	deposition
Mn	0.01	deposition
Co	0.02	Co-oxide formation of $SrSO_4$ And SiO_2 deposits
Sr, Si	003	Anode blinding via the formation
Hg	0.04	Formation of calomel
Ti hydrides	–	Formation of Ti
Fe	–	Formation of Fe oxides
TOC	5–10	Anode deactivation
F	1	Attacks Ti to form soluble

Source: R. H. Stevens, Creative Commons License.

The reduction in mol% of RuO_2 in the outermost layers of the coating can be attributed to the dissolution of RuO_2 during various processes, namely chlorine evolution, shutdowns, and oxygen generation (Keating & Behling, 1990).

The process of chlorine evolution reaction may result in the depletion of RuO_2 from the anode due to the detachment of unstable adsorbed intermediaries or oxidized surface oxides that develop during the discharge of chloride ions (Silva et al., 2009). The reaction schemes below illustrate this phenomenon.

$$RuO_2 + Cl^- \rightarrow RuO_2Cl + e$$

$$RuO_2Cl^- + H_2O + Cl^- \rightarrow RuO_2OH + Cl_2 + H^+$$

$$RuO_2OH + H^+ + e \rightarrow RuO_2 + H_2O$$

$$RuO_2 + 2H_2O \rightarrow RuO_4 + 4H^+ + 4e$$

$$RuO_2 + 5Cl^- + 3H^+ \rightarrow RuOHCl_5^{2-} + H_2O$$

Figure 4.9 illustrates the depletion of ruthenium resulting from the erosion and dissolution of adsorbed intermediates formed during the chlorine evolution reaction and surface oxides (Zhao et al., 2020). The findings indicate that the initial dissolution rate of Ru is considerably elevated but experiences a swift decline over time.

$$V_c = 2.5 \times 10^{-6} t^{-1.2}$$

The dissolution of RuO_2 during shutdowns is also evidenced in Figure 4.10. The literature lacks a clear explanation for the increased losses of RuO_2 observed during this process. One possible factor contributing to the losses of RuO_2 is the occurrence of reverse currents resulting from the cathodic reduction of sodium hypochlorite in the anolyte (Ogata et al., 1989). This phenomenon can cause the anode to become cathodic and lead to the reduction of RuO_2 to Ru or a lower oxide (e.g., RuOx, where $x < 2$), which is subsequently oxidized during electrolysis to form a soluble Ru species such as H_2RuO_3, H_2RuO_5, or RuO_2. Alternatively, the modification of the surface oxidation state of RuO_2 can occur through current interruptions, resulting in the formation of an unstable intermediate and subsequent dissolution in the anolyte. The chemical reactions that result in the dissolution of Ru exhibit a close adherence to the equations above. The observed increase in Ru losses during current interruptions can be attributed to alterations in the surface oxide composition resulting from the cathodic and anodic pulses (Momose et al., 1991). Specifically, the reverse currents associated with the cathodic pulse and the resumption of electrolysis following the anodic pulse led to the dissolution of the surface oxide at a more rapid rate than the thermally formed original oxide.

The exhaustion of RuO_2 may also ensue from the production of oxygen. The results indicate that the concentration of Ru sites on the anode surface, as evidenced by the system's overall capacitance, exhibited an initial rise followed by a decline upon anode potential escalation to unsatisfactory levels (*refer* Figure 4.11) (Eversloh et al., 2015).

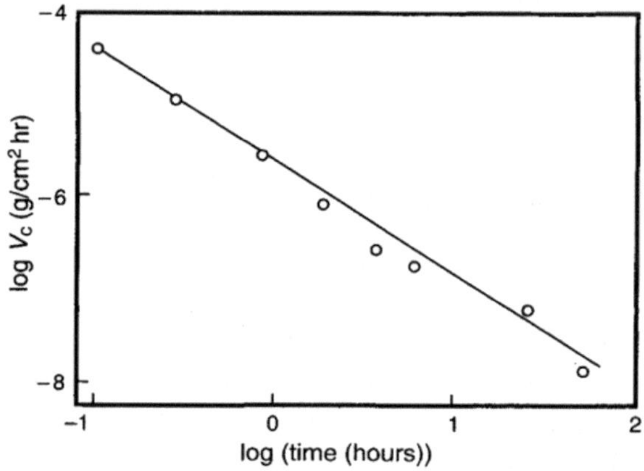

Figure 4.9. *An analysis of ruthenium depletion resulting from erosion and general dissolution. The experiment was conducted using a 30% + 70% Ti electrode at a current of 2 kA in a solution of 300 GPL NaCl at a temperature of 70°C. The results illustrate the extent of ruthenium losses under these conditions.*

Source: C. L. Hoover, Creative Commons License.

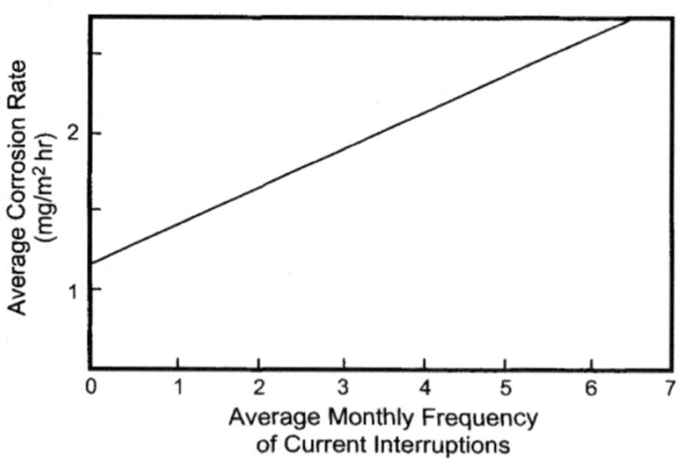

Figure 4.10. *An illustration of the correlation between the corrosion rate and the frequency of shutdowns for a +70% Ti anode.*
Source: C. L. Hoover, Creative Commons License.

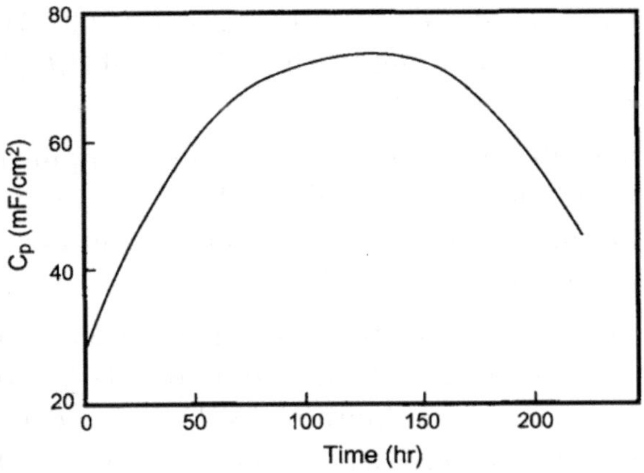

Figure 4.11. *Illustration of the pseudo capacitance's time dependency reflects anode failure.*

Source: Robert, Creative Commons License.

The observed rise in RuO_2 sites can be attributed to the emergence of surface pores, while the subsequent decline can be ascribed to the exhaustion of Ru sites (Karthikeyan et al., 2021). Upon the formation of pores on the surface, the process of electrolysis of NaCl persists within the pores. Eventually, when the chloride ions are fully consumed, the pore transitions to O_2 evolution. Previous studies have demonstrated that the RuO_2 OH and RuO_3 species mediate the process of oxygen evolution from a RuO_2 surface:

$$RuO_2 + H_2O \rightarrow RuO_4 \cdot OH + H^+ + e$$

$$RuO_2 \cdot OH \rightarrow RuO_3 + H^+ + e$$

$$RuO_3 \rightarrow RuO_2 + 1/2O_2$$

The concentration of oxygen at the anode typically exhibits (Havranek & Wippermann, 2004):

- An upward trend with an increase in the percentage of RuO_2 in the coating;
- An increase in the surface area of the coating;
- A decrease in current density;
- An increase in pH (although it remains unaffected by pH values below 2);
- Conversely, the oxygen concentration at the anode displays a downward trend with an increase in the concentration of Cl

ions (with the recommended level being 280–315 g/L) and the incorporation of dopants such as SnO_2 and HfO_2 into the coating (Jamnik & Maier, 2001).

The escalation of anode potential resulting from impurities and Ru loadings below 2 g m^{-2} is readily understandable. Nonetheless, the deactivation of the anode is noticeable when the loadings of Ru exceed 2 g m^{-2}. It is challenging to justify or explain logically.

Several recent reviews and publications have focused on the mechanisms involved in the deactivation of anodes based on RuO_2/TiO_2 during the chlorine and oxygen evolution reactions (Conway, 1991). It is widely acknowledged that, in the process of electrolysis, a certain amount of RuO_2 undergoes electrochemical dissolution (although the exact rate of dissolution remains uncertain), and a portion of it is also lost due to erosion. The RuO_2 exhibits a negligible corrosion rate at a pH range of 1 to 2 The corrosion rate of increases below a pH of 0.2 (forming RuOH) and above a pH of 4 (forming Ru and gaseous. The rate of dissolution of RuO_2 is observed to escalate with an increase in current density, a rise in % O_2 generations, a decrease in NaCl concentration, and an increase in the number of shutdowns of operating cells (Primdahl & Mogensen, 1997). It is postulated that the dissolution of TiO_2 is chemical, leading to the creation of TiO^{2+}.

The dissolution of RuO_2 coatings in NaCl and H_2SO_4 solutions during operation is widely acknowledged (Krauskopf et al., 2019). However, the correlation between this loss and the deactivation mechanism of RuO_2/TiOt anodes remains ambiguous. Several studies have suggested that the deactivation can be attributed to the surface depletion of RuO_2, which leads to a high resistivity composition of $RuO_2 + TiO_2$ when the Ru content is less than 20 atomic percent.

4.7.2. Strategies to Prevent the Loss of Ru

As previously demonstrated, reducing the proportion of RuO_2 within the external coating region results in the acquisition of unfavorable anode potential levels. The ultimate variable that leads to the accumulation of localized regions of elevated electrical current can be alleviated by employing high-quality brine that is free from impurities such as SiO_2, SiO_3^{2-}, Mn^{2+}, Ba^{2+}, Fe^{2+}, or Fe^{3+}. The mitigation of Ru losses caused by erosion resulting from chlorine gas production at the anode can be accomplished by implementing low current density operation or utilizing anode designs that alleviate gas velocities (Yu et al., 2022). However, reducing the current density is not viable as it contradicts the current convention of operating the cells at their maximum current density capacity. Reducing Ru loss during shutdowns can be accomplished by implementing various strategies, including minimizing the frequency of shutdowns, utilizing cathodic protection, or promptly eliminating hypochlorite from the cells. The maintenance of the anode coating's surface compositional integrity would be guaranteed.

Additionally, the functioning of cells within a pH spectrum of 2 to 3 would provide dual benefits. At the outset, the electrolysis of TiO_2 in an acidic environment would dissolve the substance, which would have an insignificant impact on the surface composition of the coating. Under acidic conditions, a decrease in the quantity of oxygen produced at the anode and a reduction in the loss of ruthenium resulting from the oxygen evolution reaction would be observed (Fabičovicová et al., 2015).

One potential strategy for reducing Ru losses involves the integration of IrO_2 dopants into the anode coatings. Empirical evidence has shown that the corrosion rate of ruthenium experiences a noteworthy decrease when exposed to electrolysis in Sodium Chloride solutions.

4.7.3. Rejuvenation of Deactivated RuO_2 Anodes

Assuming that the Ru loss in the deactivated anode is attributed to a uniform dissolution across the coating layer, resulting in a Ru loading that is less than 2 g m^{-2}. Under such circumstances, it becomes imperative to reapply the coating to reinstate the electrocatalytic functionality of the anode. Under the present circumstances, it is imperative to eliminate the remaining coating before applying a fresh layer (Ru et al., 2023). Assuming that the increase in anode potential is a result of the depletion of ruthenium on the surface. Therefore, the deactivated anodes' revitalization is expected to occur upon the reinstatement of these specific surface sites.

Inactivated anodes have been the subject of recent research, which has demonstrated that the utilization of electroless or electrolytic platinum deposition on flawed anodes has decreased the polarization behavior of said anodes. Furthermore, it has been noted that the longevity of said anodes is comparable to that of pristine anodes in accelerated life assessments carried out in solutions containing sulfuric acid (Cheng et al., 2022).

4.8. ANODE STRUCTURES

During the 1960s, the mercury cell was the preferred technology for chlorine production, primarily due to its capacity to function at elevated current densities. As a result, the initial commercial utilization of DSAs occurred in mercury cells, where the prompt discharge of gas was of utmost significance. The anode configurations employed in mercury cells were devised to facilitate the expeditious liberation of chlorine, thereby reducing the anode-cathode distance and maintaining high levels of NaCl concentration in this region. It has been asserted that baffles located on the rear of the active anode surface can generate adequate gas lift to propel the brine into the inter-electrode gap between the anode and cathode.

Alternative varieties of anodes for mercury cells employ slender titanium rods or delicate titanium blades. The cellular electrical linkage is established via the utilization of a boss, employing solid copper rods that are safeguarded by a titanium riser tube. Figure 4.12 depicts a conventional configuration of anodes within a mercury cell.

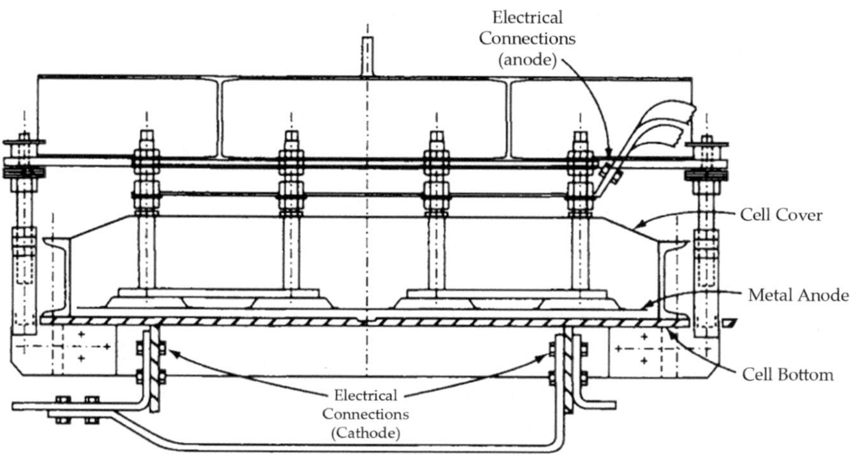

Figure 4.12. *Illustration of a mercury cell's end view showing a common configuration of metal anodes.*

Source: Tilak V. Bommaraju, Creative Commons License.

Diaphragm cells employ two fundamental anode designs: box anodes and expandable anodes. The diagram depicted in Figure 4.13 showcases the box anode specifically engineered to serve as a viable substitute for graphite anodes. The active surface comprises an expanded titanium mesh that has been coated and a titanium conductor bar encased in copper that facilitates the current flow. The anode that can be expanded, as depicted in Figure 4.14, is equipped with clips that secure the anodes before cell assembly, thereby preventing any harm to the asbestos diaphragm. Following the placement of the cathode and subsequent application of 3 mm diameter spacers over it, the anode clips are detached. Upon activation by spring, the anode surfaces converge, resulting in a consistent and narrow gap between the anode and cathode (Huang et al., 2014).

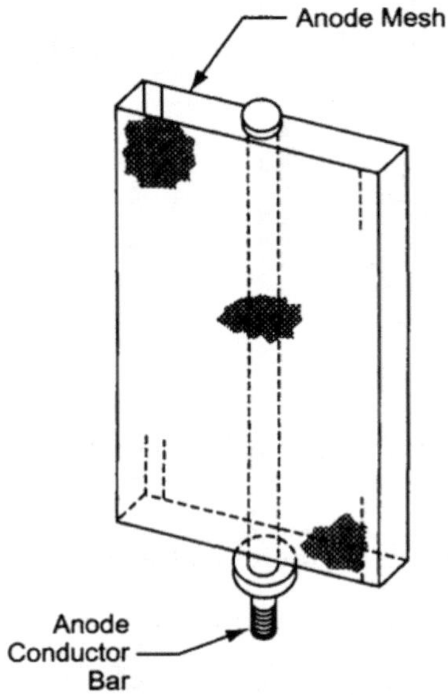

Figure 4.13. *Illustration of typical diaphragm cell box anode.*

Source: C. E. Vallet, Creative Commons License.

4.9. ANODE COSTS AND MANUFACTURERS

The pricing of an anode is contingent upon the anode's arrangement and the coating methodology and composition. The computation of the expenses incurred in applying coatings can be readily derived from the costs of labor and chemicals utilized. Typically, the cost of noble metal utilized in thermal decomposition techniques for coating falls within 30 times the original cost or \$75–100 g^{-1} of noble metal (Schmuch et al., 2018). The multiplication factor's variability stems from factors such as the anode base's inherent characteristics, the degree of repair required, the cost of labor, and the coating's applicability (Tables 4.4 and 4.5).

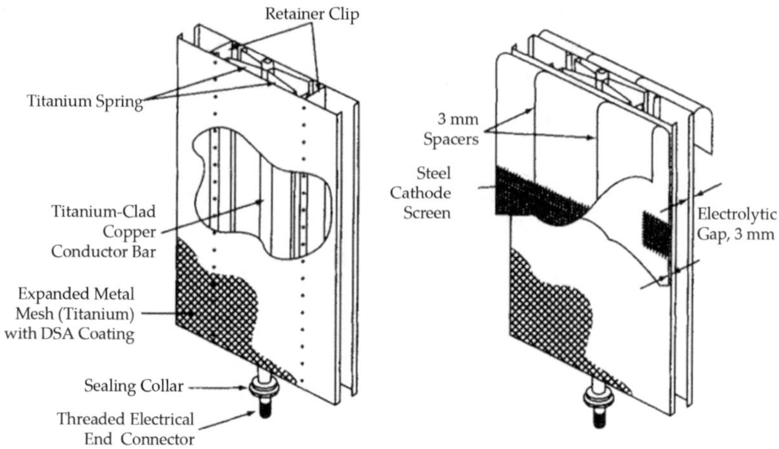

Figure 4.14. *Diaphragm cell expandable anode illustration either clamped or expanded.*

Source: J. V. Biggers, Creative Commons License.

Table 4.4. Representation of Average Lifetimes of -based Anodes

Cell Type	Average Current Density ()	Coating Composition	Ru Loading ()	Typical Life (Years)
Mercury	10		10–15	1.5–2.5 (or 180 tons)
Diaphragm	2		4–7 (mole ratio 3Ru:2Sn: 11Ti)	10 (or >200 tons)
Membrane	4		4–5 (Ru:Ir = equimolar or equal weight)	7 (expected; or 300 tons)

Source: R. Hutchings, Creative Commons License.

Figure 4.15 shows a regular anode coating procedure.

Table 4.5. Representation of ELTECH's Anodes for Chlor-Alkali Service

Designation	Primary Components	Applications
EC 100	Ru, Ti	EC 100
EC 107	Ru, Ti	EC 107
EC 150	Ru, Ti	EC 150

EC 200	Ru, Ti, Sn	EC 200
EC 300	Ru, Ti, Ir	EC 300
EC 301	Ru, Ti, Ir	EC 301
800 Series	Proprietary	Customer specified formulations

Source: T. Mussini, Creative Commons License.

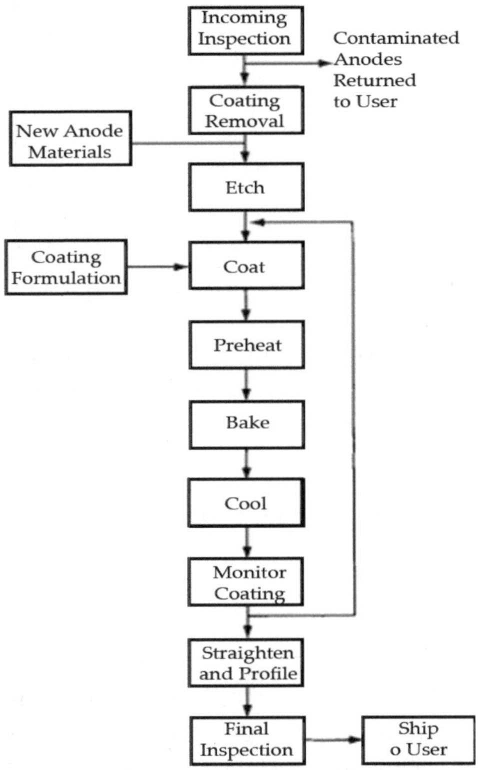

Figure 4.15. *Process flow diagram for coating metal anodes.*

Source: J. V. Biggers, Creative Commons License.

4.10. ALTERNATE ANODE COMPOSITIONS

Various anode coatings have been investigated for their potential application in chlor-alkali cells beyond those enumerated in Table 4.4. The compositions of Pt-Ir on titanium and cobalt spinels with PdO-based coatings are noteworthy (Sammes et al., 2005).

As the authors understand, these coatings have been restricted somewhat. The Ru-Ti-Ir composition has shown potential in membrane cell applications due to the addition of iridium oxide, which effectively minimizes corrosion during operation and shutdowns (as depicted in Figure 4.16).

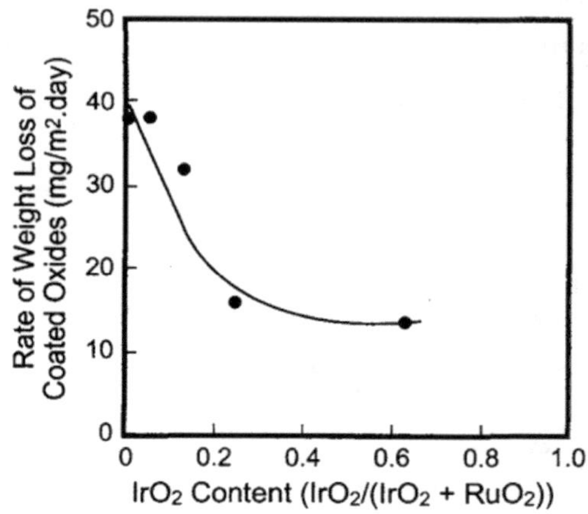

Figure 4.16. *Anode coatings with different contents are shown in terms of weight reduction. Electrolysis conditions include 210 gp[l] NaCl at 90°C, pH = 3.0, and a current density of 20 kA. Anode: 50 mottoes at 10 g produced at a firing temperature of 400°C.*

Source: R. Kotz, Creative Commons License.

4.11. SUMMARY

In the chlor-alkali industry, anodes play a crucial role in the electrolysis process. Anodes used in chlor-alkali industry are typically made of a metal substrate, such as titanium or steel, coated with a layer of a mixed metal oxide (MMO) catalyst. During the electrolysis process, the anodes facilitate the oxidation reaction at the anode where the chloride ions are oxidized to chlorine gas. Anodes in the chlor-alkali industry are subject to degradation over time due to the harsh operating conditions. Overall, anodes are vital components in the chlor-alkali industry, facilitating the production of chlorine gas and contributing to the overall electrolysis process.

REFERENCES

1. Afrasiabi, N., & Shahbazali, E., (2011). RO brine treatment and disposal methods. *Desalination and Water Treatment, 35*(1–3), 39–53.
2. Ammar, A. U., Stefan, M., Macavei, S. G., Tripon, S., Pana, O., Leostean, C., & Erdem, E., (2023). Characterization of defect structures in nanoscaled W-doped TiO_2 tested as supercapacitor electrode materials. *Journal of Materials Science: Materials in Electronics, 34*(2), 1–18.
3. Arunkumar, M., & Paul, A., (2017). Importance of electrode preparation methodologies in supercapacitor applications. *ACS Omega, 2*(11), 8039.
4. Battisti, A. D., Lodi, G., Nanni, L., Battaglin, G., & Benedetti, A., (1997). Preparation and characterization of oxide film electrodes. *Canadian Journal of Chemistry, 75*(11), 1759–1765.
5. Bauerfeind, K. C., Laun, J., Frisch, M., Kraehnert, R., & Bredow, T., (2022). Metal substitution in rutile TiO_2: Segregation energy and conductivity. *Journal of Electronic Materials*, 1–12.
6. Betelu, S., Polychronopoulou, K., Rebholz, C., & Ignatiadis, I., (2011). Novel CeO_2-based screen-printed potentiometric electrodes for pH monitoring. *Talanta, 87*, 126–135.
7. Bommaraju, T. V., Chen, C. P., & Birss, V. I., (2008). Deactivation of thermally formed RuO_2+ TiO_2 coatings during chlorine evolution: Mechanisms and reactivation measures. In: Moorhouse, J., (ed.), *Modern Chlor-Alkali Technology* (pp. 57–81).
8. Brightman, E., Ivey, D. G., Brett, D. J. L., & Brandon, N. P., (2011). The effect of current density on H_2S-poisoning of nickel-based solid oxide fuel cell anodes. *Journal of Power Sources, 196*(17), 7182–7187.
9. Burke, L. D., Mulcahy, J. K., & Whelan, D. P., (1984). Preparation of an oxidized iridium electrode and the variation of its potential with pH. *Journal of Electroanalytical Chemistry and Interfacial Electrochemistry, 163*(1–2), 117–128.
10. Carvalho, N. J. M., Zoestbergen, E., Kooi, B. J., & De Hosson, J. T. M., (2003). Stress analysis and microstructure of PVD monolayer TiN and multilayer TiN/(Ti, Al) N coatings. *Thin Solid Films, 429*(1–2), 179–189.
11. Casimir, A., Zhang, H., Ogoke, O., Amine, J. C., Lu, J., & Wu, G., (2016). Silicon-based anodes for lithium-ion batteries: Effectiveness of materials synthesis and electrode preparation. *Nano-Energy, 27*, 359–376.
12. Cheng, F., Cao, F., Ru Fan, F., & Wu, B., (2022). Promotion strategies of hole transport materials by electronic and steric controls for n-i-p perovskite solar cells. *ChemSusChem, 15*(14), e202200340.

13. Cimenti, M., & Hill, J. M., (2009). Direct utilization of liquid fuels in SOFC for portable applications: Challenges for the selection of alternative anodes. *Energies, 2*(2), 377–410.
14. Conway, B. E., (1991). Transition from "supercapacitor" to "battery" behavior in electrochemical energy storage. *Journal of the Electrochemical Society, 138*(6), 1539.
15. Evdokimov, S. V., (2001). Electrochemical and corrosion behavior of dimensionally stable anodes in chlorate electrolysis: Efficiency of the sodium chlorate production at elevated temperatures. *Russian Journal of Electrochemistry, 37*, 363–370.
16. Eversloh, C. L., Schulz, M., Wagner, M., & Ternes, T. A., (2015). Electrochemical oxidation of tramadol in low-salinity reverse osmosis concentrates using boron-doped diamond anodes. *Water Research, 72*, 293–304.
17. Fabičovicová, K., Lucas, M., & Claus, P., (2015). From microcrystalline cellulose to hard-and softwood-based feedstocks: Their hydrogenolysis to polyols over a highly efficient ruthenium–tungsten catalyst. *Green Chemistry, 17*(5), 3075–3083.
18. Faes, A., Hessler-Wyser, A., Zryd, A., & Van Herle, J., (2012). A review of RedOx cycling of solid oxide fuel cells anode. *Membranes, 2*(3), 585–664.
19. Feijoo, S., Yu, X., Kamali, M., Appels, L., & Dewil, R., (2023). Generation of oxidative radicals by advanced oxidation processes (AOPs) in wastewater treatment: A mechanistic, environmental, and economic review. *Reviews in Environmental Science and Bio/Technology, 22*(1), 205–248.
20. Geuenich, D., Hess, K., Köhler, F., & Herges, R., (2005). Anisotropy of the induced current density (ACID), a general method to quantify and visualize electronic delocalization. *Chemical Reviews, 105*(10), 3758–3772.
21. Havranek, A., & Wippermann, K., (2004). Determination of proton conductivity in anode catalyst layers of the direct methanol fuel cell (DMFC). *Journal of Electroanalytical Chemistry, 567*(2), 305–315.
22. hi, X., & Misawa, H., (2021). Plasmonic photochemical water splitting for efficient solar energy conversion. *Progress in Nanophotonics, 6*, 71–101.
23. Horvick, E. W., (2006). Zinc in the world of electroplating. *Plating and Surface Finishing, 93*(6), 42–48.
24. Huang, C., Xiao, J., Shao, Y., Zheng, J., Bennett, W. D., Lu, D., & Liu, J., (2014). Manipulating surface reactions in lithium–sulfur batteries using hybrid anode structures. *Nature Communications, 5*(1), 3015.

25. Jamnik, J., & Maier, J., (2001). Generalized equivalent circuits for mass and charge transport: Chemical capacitance and its implications. *Physical Chemistry Chemical Physics, 3*(9), 1668–1678.
26. Jiang, M., Wang, H., Li, Y., Zhang, H., Zhang, G., Lu, Z., & Jiang, L., (2017). Superaerophobic RuO_2-based nanostructured electrode for high-performance chlorine evolution reaction. *Small, 13*(4), 1602240.
27. Karlsson, R. K., & Cornell, A., (2016). Selectivity between oxygen and chlorine evolution in the chlor-alkali and chlorate processes. *Chemical Reviews, 116*(5), 2982–3028.
28. Karthikeyan, S., Narenthiran, B., Sivanantham, A., Bhatlu, L. D., & Maridurai, T., (2021). Supercapacitor: Evolution and review. *Materials Today: Proceedings, 46*, 3984–3988.
29. Keating, J. T., & Behling, K. J., (1990). Brine, impurities, and membrane chlor-alkali cell performance. *Modern Chlor-Alkali Technology, 4*, 125–139.
30. Kim, J. H., Zhu, K., Yan, Y., Perkins, C. L., & Frank, A. J., (2010). Microstructure and pseudocapacitive properties of electrodes constructed of oriented NiO–TiO_2 nanotube arrays. *Nano-Letters, 10*(10), 4099–4104.
31. Kim, T. Y., & Yang, S., (2014). Fabrication method and characterization of electrodeposited and heat-treated iridium oxide films for pH sensing. *Sensors and Actuators B: Chemical, 196*, 31–38.
32. Koutu, V., Shastri, L., & Malik, M. M., (2016). Effect of NaOH concentration on optical properties of zinc oxide nanoparticles. *Materials Science-Poland, 34*(4), 819–827.
33. Krauskopf, T., Mogwitz, B., Rosenbach, C., Zeier, W. G., & Janek, J., (2019). Diffusion limitation of lithium metal and Li–Mg alloy anodes on LLZO type solid electrolytes as a function of temperature and pressure. *Advanced Energy Materials, 9*(44), 1902568.
34. Kuo, M. C., & Yen, S. K., (2002). The process of electrochemical deposited hydroxyapatite coatings on biomedical titanium at room temperature. *Materials Science and Engineering: C, 20*(1–2), 153–160.
35. Lakshmanan, S., & Murugesan, T., (2014). The chlor-alkali process: work in progress. *Clean Technologies and Environmental Policy, 16*, 225–234.
36. Mai, A., Haanappel, V. A., Uhlenbruck, S., Tietz, F., & Stöver, D., (2005). Ferrite-based perovskites as cathode materials for anode-supported solid oxide fuel cells: Part I. Variation of composition. *Solid State Ionics, 176*(15–16), 1341–1350.
37. Maier, B. R., Garcia-Diaz, B. L., Hauch, B., Olson, L. C., Sindelar, R.

L., & Sridharan, K., (2015). Cold spray deposition of Ti_2AlC coatings for improved nuclear fuel cladding. *Journal of Nuclear Materials, 466*, 712–717.

38. Mauger, A., & Julien, C. M., (2021). Remedies to avoid failure mechanisms of lithium-metal anode in Li-ion batteries. *Inorganics, 10*(1), 5.
39. Mishra, A. K., & Ramaprabhu, S., (2011). Functionalized graphene-based nanocomposites for supercapacitor application. *The Journal of Physical Chemistry C, 115*(29), 14006–14013.
40. Mohtadi, R., Lee, W. K., & Van Zee, J. W., (2004). Assessing durability of cathodes exposed to common air impurities. *Journal of Power Sources, 138*(1–2), 216–225.
41. Momose, T., Higuchi, N., Arimoto, O., Yamaguchi, K., & Walton, C. W., (1991). Effects of low concentration levels of calcium and magnesium in the feed brine on the performance of a membrane chlor-alkali cell. *Journal of the Electrochemical Society, 138*(3), 735.
42. Monreal, M., Lafoz, E., Olive, A., Del Rio, L., & Vedia, C., (1994). Comparison of subcutaneous unfractionated heparin with a low molecular weight heparin (Fragmin®) in patients with venous thromboembolism and contraindications to coumarin. *Thrombosis and Hemostasis, 71*(01), 007–011.
43. Nayar, K. G., Fernandes, J., McGovern, R. K., Dominguez, K. P., McCance, A., & Al-Anzi, B. S., (2019). Cost and energy requirements of hybrid RO and ED brine concentration systems for salt production. *Desalination, 456*, 97–120.
44. Ogata, Y., Kojima, T., Uchiyama, S., Yasuda, M., & Hine, F., (1989). Effects of the brine impurities on the performance of the membrane-type chlor-alkali cell. *Journal of the Electrochemical Society, 136*(1), 91.
45. Olsen, E., & Thonstad, J., (1999). Nickel ferrite as inert anodes in aluminum electrolysis: Part II Material performance and long-term testing. *Journal of Applied Electrochemistry, 29*, 301–311.
46. Pilla, A. S., Cobo, E. O., Duarte, M. M. E., & Salinas, D. R., (1997). Evaluation of anode deactivation in chlor-alkali cells. *Journal of Applied Electrochemistry, 27*, 1283–1289.
47. Pouladvand, I., Asl, S. K., Hoseini, M. G., & Rezvani, M., (2019). Characterization and electrochemical behavior of $Ti/TiO_2–RuO_2–IrO_2–SnO_2$ anodes prepared by sol-gel process. *Journal of Sol-Gel Science and Technology, 89*, 553–561.
48. Preston-Thomas, H., (1990). The international temperature scale of 1990(ITS-90). *Metrologia, 27*(1), 3–10.

49. Primdahl, S., & Mogensen, M., (1997). Oxidation of hydrogen on Ni/yttria-stabilized zirconia cermet anodes. *Journal of the Electrochemical Society, 144*(10), 3409.
50. Reddy, A. L. M., & Ramaprabhu, S., (2007). Nanocrystalline metal oxides dispersed multiwalled carbon nanotubes as supercapacitor electrodes. *The Journal of Physical Chemistry C, 111*(21), 7727–7734.
51. Robinson, H. A., & George, P. F., (1954). Effect of alloying and impurity elements in magnesium alloy cast anodes. *Corrosion, 10*(6), 182–188.
52. Ru, Q., Li, Y., Xie, W., Ding, Y., Chen, L., Xu, G., & Wang, F., (2023). Fighting age-related orthopedic diseases: Focusing on ferroptosis. *Bone Research, 11*(1), 12.
53. Sammes, N. M., Du, Y., & Bove, R., (2005). Design and fabrication of a 100 W anode supported micro-tubular SOFC stack. *Journal of Power Sources, 145*(2), 428–434.
54. Schmuch, R., Wagner, R., Hörpel, G., Placke, T., & Winter, M., (2018). Performance and cost of materials for lithium-based rechargeable automotive batteries. *Nature Energy, 3*(4), 267–278.
55. Sen, U., (2004). Kinetics of titanium nitride coatings deposited by thermo-reactive deposition technique. *Vacuum, 75*(4), 339–345.
56. Silva, J. M., Soloveichik, G. L., & Novak, D., (2009). Effects of organic impurities on chloralkali membrane electrolyzer performance. *Industrial and Engineering Chemistry Research, 48*(2), 983–987.
57. Sundriyal, S., Sharma, M., Kaur, A., Mishra, S., & Deep, A., (2018). Improved electrochemical performance of rGO/TiO_2 nanosheet composite based electrode for supercapacitor applications. *Journal of Materials Science: Materials in Electronics, 29*, 12754–12764.
58. Tilak, B. V., Birss, V. I., Wang, J., Chen, C. P., & Rangarajan, S. K., (2001). Deactivation of thermally formed Ru/Ti oxide electrodes: an AC impedance characterization study. *Journal of the Electrochemical Society, 148*(9), D112.
59. Tilak, B. V., Chen, C. P., Birss, V. I., & Wang, J., (1997). Capacitive and kinetic characteristics of Ru–Ti oxide electrodes: Influence of variation in the Ru content. *Canadian Journal of Chemistry, 75*(11), 1773–1782.
60. Tong, S. P., Ma, C. A., & Feng, H., (2008). A novel PbO_2 electrode preparation and its application in organic degradation. *Electrochimica Acta, 53*(6), 3002–3006.
61. Uribe, F. A., Gottesfeld, S., & Zawodzinski, T. A., (2002). Effect of ammonia as potential fuel impurity on proton exchange membrane fuel

cell performance. *Journal of the Electrochemical Society, 149*(3), A293.

62. Wang, H., Xie, Z., Liu, C., Hu, B., Liao, S., Yan, X., & Ouyang, C., (2023). Rate-dependent failure mechanisms and mitigating strategies of anode-free lithium metal batteries. *ACS Applied Materials and Interfaces, 15*(10), 12967–12975.
63. Wang, J., Zhai, C., Chen, H., Cong, N., Tan, L., Fang, H., & Zhu, Y., (2019). Preparation of the ordered spherical Ru-RuO_2 and electrocatalysis toward chlorine evolution reaction. *Int. J. Electrochem. Sci., 14*, 10187–10197.
64. Wang, Y., Liu, Y., Wiley, D., Zhao, S., & Tang, Z., (2021). Recent advances in electrocatalytic chloride oxidation for chlorine gas production. *Journal of Materials Chemistry A, 9*(35), 18974–18993.
65. Yao, S., Wang, M., & Madou, M., (2001). A pH electrode based on melt-oxidized iridium oxide. *Journal of the Electrochemical Society, 148*(4), H29.
66. Yokokawa, H., Yamaji, K., Brito, M. E., Kishimoto, H., & Horita, T., (2011). General considerations on degradation of solid oxide fuel cell anodes and cathodes due to impurities in gases. *Journal of Power Sources, 196*(17), 7070–7075.
67. Yu, J., Wang, X., Huang, X., Cao, J., Huo, Q., Mi, L., & He, C., (2022). Confining ultrafine Ru clusters into TiO_2 lattice frameworks to yield efficient and ultrastable electrocatalysts towards practical hydrogen evolution. *Chemical Engineering Journal, 446*, 137248.
68. Zdrachek, E., Forrest, T., & Bakker, E., (2023). Symmetric cell for improving solid-contact pH electrodes. *Analytica Chimica Acta, 1239*, 340652.
69. Zhao, Z., Liu, G., Jia, H., & He, L., (2020). Sandwiched liquid-membrane electrodialysis: Lithium selective recovery from salt lake brines with high Mg/Li ratio. *Journal of Membrane Science, 596*, 117685.
70. Zhu, Z., Liu, X., Ye, Z., Zhang, J., Cao, F., & Zhang, J., (2018). A fabrication of iridium oxide film pH micro-sensor on Pt ultramicroelectrode and its application on *in-situ* pH distribution of 316L stainless steel corrosion at open circuit potential. *Sensors and Actuators B: Chemical, 255*, 1974–1982.

CHAPTER

5 Cathodes in Chlor-Alkali Industry

Contents

5.1. INTRODUCTION

The principal electrochemical process transpiring at the cathode within the diaphragm along with membrane cells seems to be the generation of hydrogen via alkaline solutions. In diaphragmatic cells, the cathode material typically comprises carbon steel, whereas in membrane cells, it is commonly composed of nickel. Upon analyzing the constituents of cell voltage within these cells, it has been observed that the cathodes overvoltage falls within the bracket of 0.2–0.3 V within the current density ranging from 2–4 kA on steel substrates and exhibits a slightly elevated value on nickel substrates (Moussallem et al., 2008; Li et al., 2021; Furuya & Aikawa, 2000; Chatenet et al., 2003).

Implementing metallic anodes was a significant accomplishment within the chlor-alkali sector during the beginning of the 1960s. The replacement of carbon anodes alongside precious metal-infused anodes yielded a reduction in voltage of more than 0.2 V, an increase in anode lifespan, and improved cell gas (i.e., greater current efficiency). It enabled the cells to operate at higher loads without substantial voltage penalties. The global adoption of DSA anodes for commercial purposes has stimulated the water electrolysis sector to explore alternative materials, including mixed oxides for anodes as well as high surface areas of nickel-based electrodes for cathodes, with the aim of minimizing energy consumption. The discrepancy between the voltage at which the cathode responsible for hydrogen production operates and the potential for reversible hydrogen release in the specific electrolyte is referred to as the hydrogen excessive voltage generated by the cathode (Moussallem et al., 2012; Lipp et al., 2005). The Tafel formula is typically followed by hydrogen excessive voltage within industrial settings:

$$n = b \ln\left(\frac{I}{i_0 A}\right)$$

The variables above are defined as follows: q represents the hydrogen excessive voltage, b denotes the Tafel slope, A signifies the overall surface area, represents the interaction current density measured in amps per centimeter squared of the actual surface area, as well as i_o denotes the total current (where $I = iA$, i represents the actual current density within amps per square centimeters for the electrochemically engaged surface area). The values of along with b are contingent upon the HER pace on the specific substrate as well as the HER mechanism. On shiny metallic surfaces within alkaline solutions at 25°C, the gradient of Tafel b is estimated to be around 120 mV (Ichinose et al., 2004; Jović et al., 2014). The significant overvoltage observed during hydrogen release on mercury renders the concept of a chlorine cell built around a mercury cathode feasible. In conventional mercury cells, the potential of the cathode exhibits a cathodic shift of approximately 1.8 V relative to the potential of the reversed hydrogen electrode. A typical sample of chlorine gas derived from a cell may comprise 0.5% , thereby signifying that the ion made up of hydrogen carries 0.59% of the overall cathodic current, while the sodium ion carries the remaining 99.5%. The overvoltage resulting from sodium on the cathode made of mercury is relatively negligible, approximately at the magnitude of 10^{-20} millivolts (Lakshmanan & Murugesan, 2014).

The conventional cathode substance utilized in diaphragm cells seems low carbon steel. This material showcases a low hydrogen excessive voltage as well as excellent endurance in saltwater solutions. Additionally, it is easily accessible at affordable rates in a diverse range of structural forms. The process involves cathodic protection to prevent corrosion throughout the operation, along with appropriate maintenance; it yields a production rate of 256–300 tons for NaOH per square meter of cathode area. Steel has been found to be a comparatively effective electromagnetic catalyst for the creation of hydrogen. However, in diaphragmatic chlorine cells, it exhibits an overvoltage of approximately 300 mV, contingent upon the condition of the cathode's outermost layer (as depicted in Figure 5.1). In addition, it can be observed that iron dissolves in the form of ferrate, HFe when subjected to hot and concentrated caustic substances at an open-circuit potential. This observation is consistent with the predictions made by the Pourbaix illustration in Figure 5.2. This scenario can also occur in instances where the cell experiences a short circuit over shutdowns. This is due to the sudden polarization of the cathode's outermost layer, which is devoid of any oxide, within an anodic manner (Chavan et al., 2015; O'Brien et al., 2005).

Did You Know?

Charles Watt, along with Charles Burgess Hall, created the mercury cell technique in 1892 (Vorce, 1944).

In contrast to iron, nickel displays lower electrocatalytic activity toward the hydrogen evolution reaction (HER). However, it demonstrates remarkable resistance to corrosion when exposed to highly concentrated and alkaline solutions at elevated temperatures. The endurance of Ni within caustic solutions (as shown in Figure 5.3) and the research conducted by the fluid electrolysis sector for generating Ni-based cathodes have inspired considerable endeavors toward the development of catalytic cathodes suitable for use in chlor-alkali cells (Iwakura et al., 1995; Jaccaud et al., 1989).

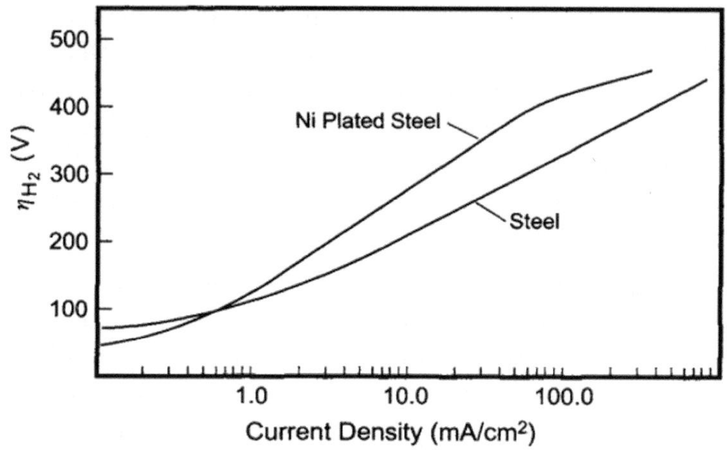

Figure 5.1. *Representation of polarization data for the HER on steel and Ni plated steel in 15% NaOH +17% NaCl solutions at 95°C.*

Source: Tilak et al. (1986).

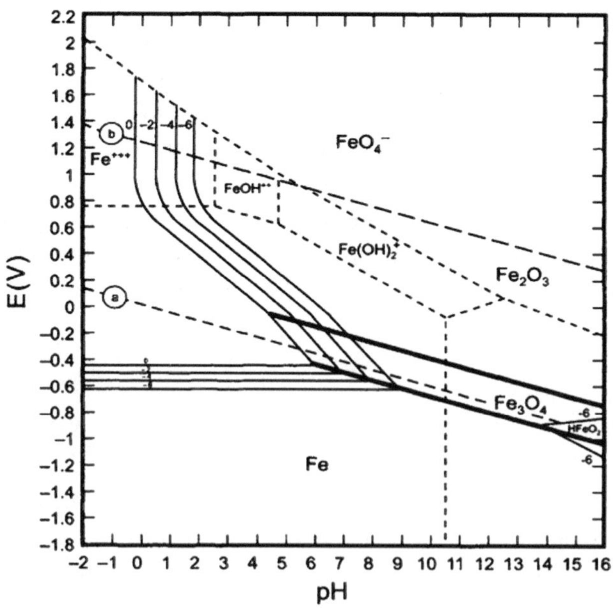

Figure 5.2. *Representation of potential-pH diagram for the Fe-0 system at 25°C. The region between the lines (a) and (b) represents the domain of the*

thermodynamic stability of water. The numbers on the lines represent the logarithm of the activity of the species.

Source: J. D. Portar, Creative Commons License.

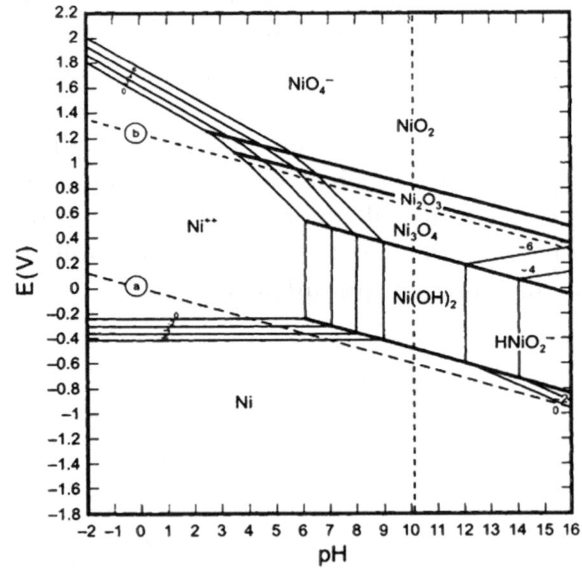

Figure 5.3. *Potential-pH diagram for the Fe-0 system at 25°C. The region between the lines (a) and (b) represents the domain of the thermodynamic stability of water. The numbers on the lines represent the logarithm of the activity of the species.*

Source: J. D. Portar, Creative Commons License.

5.2. CRITERIA FOR LOW OVERVOLTAGE HYDROGEN CATHODE

The approach towards achieving materials with superior performance over hydrogen release in highly alkaline environments has been two-fold:

- Selection of materials with high it and low b;
- The augmentation of the outermost area.

Materials that have elevated , as well as low b, have been chosen due to their ability to achieve high-functioning current densities lacking a significant increase within overvoltage (Endoh et al., 1988; Zhao et al., 2019; Giz et al., 2003), as depicted in Figure 5.4. The HER can be analyzed kinetically by considering the commonly acknowledged pathways. It has been observed that the involvement of adsorbed intermediaries in the rate-determining step leads to low Tafel slopes (as presented in Table 5.1).

The diffusion of atomic hydrogen associated with the slow step towards the mass of the metal is a characteristic of metals with low Tafel slopes. These metals serve as retained hydrogen sinks, which can be beneficial in certain scenarios, like buffering opposite currents throughout shutdowns. However, it can also adversely affect the structural strength of the metal. Therefore, the optimal material would possess a high initial exchange current density (i_0) for the HER, where the rate-limiting step seems to be the ejection process. One non-trivial factor that is apparent is the augmentation of the region, leading to a reduction in the excessive voltage for hydrogen, eventually ending in hydrogen embrittlement.

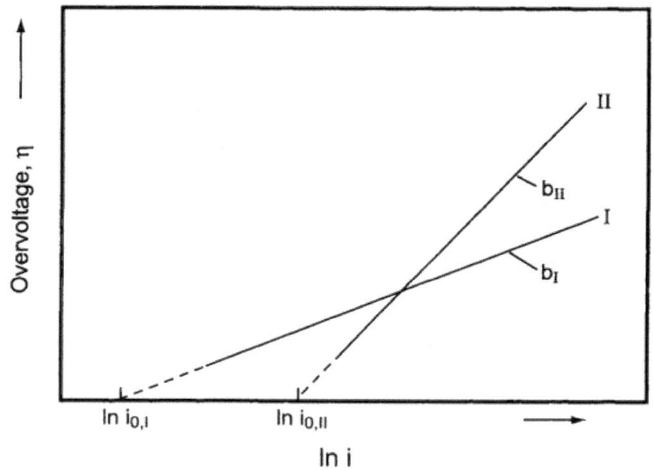

Figure 5.4. *Schematic of the polarization data when two activation-controlled reactions with different Tafel slopes proceed.*

Source: B. E. Conway, Creative Commons License.

Table 5.1. Pathways Involved in the H_2 Evolution Reaction

Mechanism	**Fd**	**Comments**
Discharge M + + e MH	2	Non-embrittling.
Recombination MH + MH M +	0.5 (1)	Hydrogen embrittlement possible.

Electrochemical desorption MH + + e 2M +	1/(1.5)	Hydrogen embrittlement possible.

Source: Fumio Hine, Creative Commons License.

The primary technical considerations for choosing an efficient cathode material encompass the following:

- Cost/life;
- Electrocatalytic activity;
- Stability when the circuit is open;
- Coating process in relation to cathode geometry and cell architecture;
- Product purity.

5.2.1. Cost Life

Assuming a catalytic cathode yield a reduction of 100 kilowatts one-hour for chlorine, the cost saving can be calculated. At an operational current intensity of 3 kA , an electricity cost of \$0.035 kW I, and 100% efficiency, the estimated cost savings throughout a period of five years, with 360 operational days every year would amount to \$600. With a cathode coating expense of \$ 600 the time required to recoup the initial investment is five years (Hachiya, 2009; Jung et al., 2014). The calculation of a payback period T can be determined by utilizing the subsequent equation.

T=31.5P / LEDSC

where; T is the payback time; L is the operating load, kA ; E is the current efficiency, fraction; D is the online factor, days; S is the energy saving, kW ; C is the cost of DC power, \$-kW ; and P is the coating cost,.

This concise equation demonstrates that the cost seems contingent upon various factors, including but not limited to current density, energy conservation, power expenses, and operational longevity, with the additional consideration of potential expenses incurred during periods of cathode replacement or maintenance (Joudaki et al., 2010; Garcia-Herrero et al., 2017; Guo et al., 2020).

5.2.2. Electro Catalytic Activity

As previously mentioned, there is a lack of straightforward predictive methods for determining the optimal cathode structure that can yield minimal overvoltages within alkaline solutions. The swapped current density, as well as the Tafel slope over the HER on different metals, are presented in the study (Garcia-Herrero et al., 2017; Antozzi et al., 2008). Additionally, Figures 5.5 and 5.6 illustrate the cathodic polarization spectrum of select metals in solutions containing NaOH and NaCl at a temperature of 95°C. The data collected from planar surfaces indicate that the hydrogen excessive potential for all the aforementioned metals falls within a spectrum of 300–350 mV during 200–300 mA , comparable with the hydrogen excessive voltage observed in steel (Hou et al., 2018).

Due to its high stability in alkaline solutions, nickel has been a subject of considerable attention in the development of catalytic properties through the process of alloying with both metallic and non-metallic elements. The underlying principle of this methodology involves the potential for eliciting electrocatalytic activity in nickel with respect to the HER, through the integration of metals situated on both sides of the volcano plot, which is built on the basis of the thermal effects of forming metal-hydrogen bonds, within the nickel matrix. The impact of alloying has been demonstrated through experimentation with diverse nickel alloys along with additional compounds, as depicted in Figures 5.7 and 5.8. The activity that exhibits optimal performance from a composition comprising 10% Co and 90% Ni is depicted in Figure 5.7. The Ni-Sn alloy demonstrated properties that were more akin to those of nickel. However, NiTi, along with Ni, displayed significant effects by manifesting low overvoltages within the region of a high density of current. The metals, however, underwent disintegration within a few hours due to the creation of titanium hydrides and the consequent expansion within the lattice. The alloys of Ni–Ir, Ni–W, as well as Ni–Co displayed elevated overpotentials in comparison to the ones currently utilized in the industrial sector (Zhang et al., 2021; Figueiredo et al., 2013; Salah et al., 2022; Kiros et al., 2006).

Figure 5.8 illustrates the impact of B, P, Bi, S, SiC, Si, as well as Zron the current-voltage characteristics within the nickel matrix. The observed increase in catalytic efficiency was noteworthy. However, the overvoltages remained elevated with the exception of NiSi and . The results indicated elevated o values, were defined as A, and reduced Tafel slopes. The heightened level of activity observed can be attributed to the increased surface area resulting from the extraction of silicon from the NiSi as well as the Ni matrix. The comparison of the low Tafel slopes ranging from 20–30 mV compared with the Tafel gradient of approximately 110 mV detected in the uniform nickel surfaces seems an intriguing observation. The identification of minimal over-voltages prompted comprehensive research endeavors aimed at developing nickel-based wires possessing substantial surface areas (Roh et al., 2019).

Various methods are at disposal for producing extensive surface areas, among which plasma or just thermal spraying process, thermal breakdown, as well as electro-deposition are among the most commonly employed and feasible. The processes as mentioned above, have the capability to generate nickel coatings that are of high purity, as well as composites that comprise nickel, as well as an element that can be leached in caustic solutions, such as Zn, Al, S, Mo, and Si (Nickell & Tatarchuk, 2005). The diverse composite materials that were analyzed are being discussed.

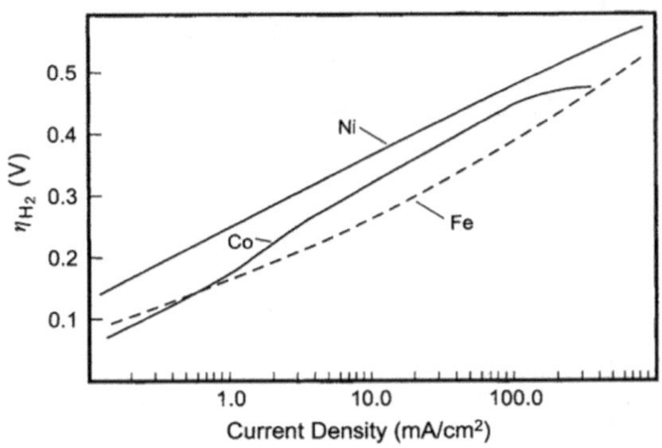

Figure 5.5. *Illustration of polarization data for the HER on Fe, Co, and Ni electrodes in 15% NaOH+ 17% NaCl solutions at 95°C.*

Source: H. Kita, Creative Commons License.

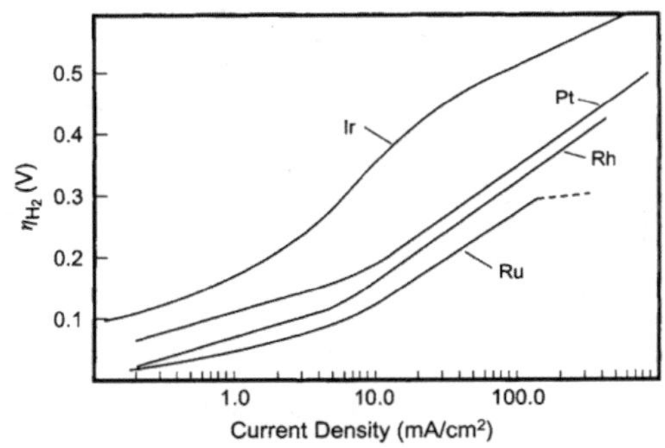

Figure 5.6. *Illustration of polarization data for the HER on Pt, Ir, Rh, and Ru electrodes in 15% NaOH + 17% NaCl solutions at 95°C.*

Source: H. Kita, Creative Commons License.

5.2.3. Stability Under Open Circuit Conditions

The majority of metals exhibit stability under cathodic protection. Nevertheless, a considerable number of them exhibit instability when subjected to extremely diluted caustic solutions at elevated temperatures. Iron cathodes immersed in a NaOH solution demonstrate an adverse open-circuit value in comparison to the reversed hydrogen electrodes present in that exact solution.

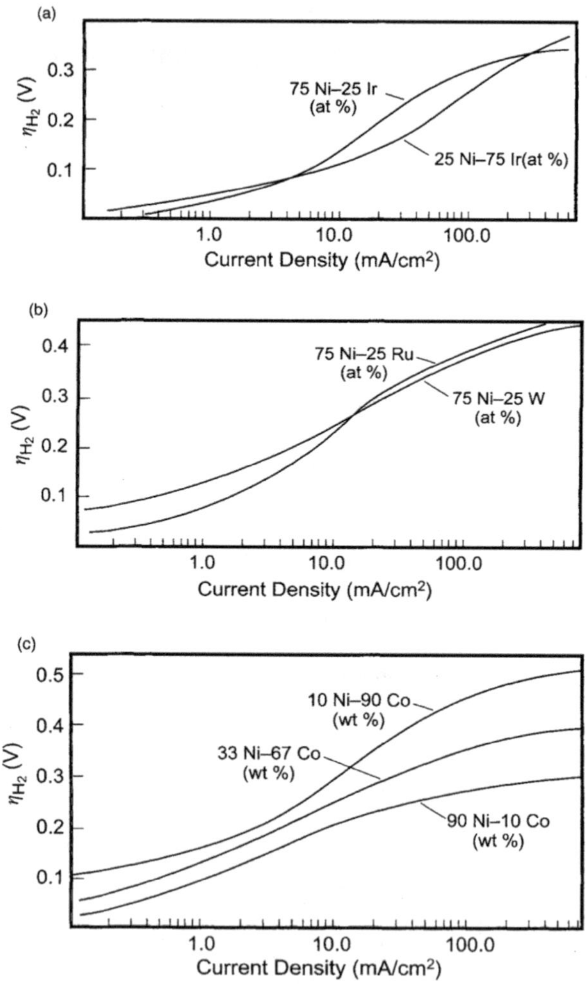

Figure 5.7. *Illustration of polarization data for the HER on (a) Ni–Ir alloys; (b) Ni–Ru alloys; and (c) Ni–Co alloys in 15% NaOH + 17% NaCl solutions at 95°C.*

Source: H. Kita, Creative Commons License.

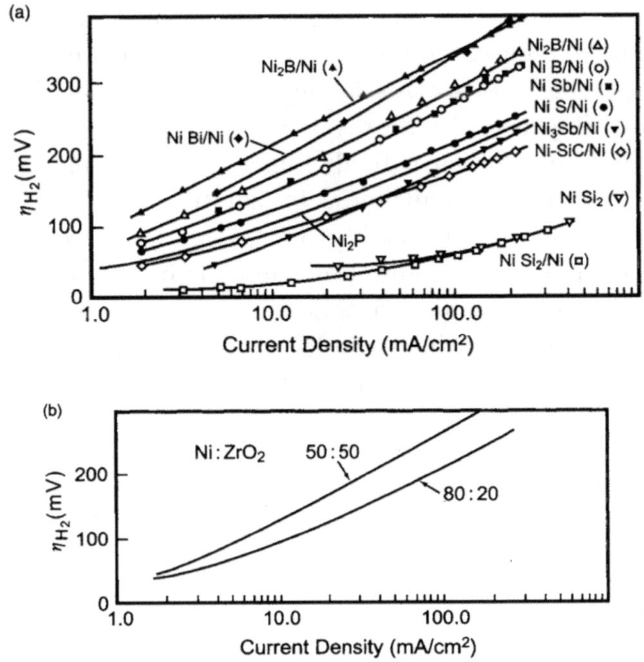

Figure 5.8. *Illustration of polarization data for the HER on (a) various Ni-based compounds; and (b) Ni-Zr mixtures plasma sprayed on Ni substrates in 15% NaOH + 17% NaCl solutions at 95°C.*

Source: R. Parsons, Creative Commons License.

A chance of corrosion is generated due to the anodized dissolution of iron in the form of HFe. The reduction process that corresponds to this phenomenon involves the ejection of molecules, leading to the formation of hydrogen. Alternatively, it may involve the elimination of distinct catholyte species, including and (Mirzazadeh et al., 2008).

The pH level of the solution can be manipulated to control the deterioration of iron that occurs because of the former. Still, the induced instability of Fe due to the cathodic decline in or is not readily preventable. Therefore, it is imperative to select a cathode material that exhibits stability in NaOH solutions with high alkalinity and can withstand oxidation caused by the decreased reactions of and (Ito et al., 2022).

5.2.4. Selection of the Coating Technique

Various techniques, including plasma spraying, heat spraying, electrodeposition, as well as thermal degradation, can be utilized to produce coatings with a large surface area upon substrates made of iron or nickel. It is imperative to take into account the operational temperature and intricacy of the cathode framework when selecting the plating technology due to the restricted "throwing power" of the diverse coating techniques. Consequently, the task of coating a complete cathode finger within the intricate diaphragm cell presents significant challenges. In addition, the fabrication limits pertaining to the cathodes are highly strict, rendering any coating method that necessitates high-temperature procedures liable to cause distortion in the cathode. The aforementioned phenomenon causes a distortion in the present cell distribution, which results in an elevation of the structural electromagnetic drop (Jorne & Louvar, 1980).

Remember:

Cathodes utilized in the chlor-alkali sector are susceptible to deterioration, necessitating regular upkeep and substitution to guarantee peak efficiency and product excellence (Perry et.al, 2020).

5.2.5. Product Purity

Chlor-alkali cells have been subject to two parasitical cathode reactions that result in "blind" current effectiveness losses:

$$OCl^- + H_2O + 2e \rightarrow Cl^- + 2OH^-$$

$$ClO_3^- + 3H_2O + 6e \rightarrow Cl^- + 6OH^-$$

It is a well-established fact that under circumstances of restricting current density, all metallic substances are capable of reducing at the cathode. However, the cathodic elimination of to is a slow process kinetically and is observed only in areas depending on iron at minimal current densities. The presence of chlorate within caustic is considered unfavorable for certain applications, thus conferring a benefit to cathodes based on iron. Nickel is preferred over iron as the metal preference over membrane-cell cathodes due to the drawbacks associated with iron. A potential objective for future advancements would be to enhance the electrocatalytic performance for the reduction of chlorate (Zhang et al., 2023).

5.3. METHODS EMPLOYED TO REALIZE HIGH SURFACE AREA

As previously mentioned, cathodes composed of metallic materials or alloys do not exhibit superior electrocatalytic properties compared to those made of steel or nickel. The achievement of low cathode excessive voltages is contingent upon the substantial cathode area. The endeavor to attain reduced cathodic overvoltages has been centered on the surface coatings of the cathode. Various methods can be employed to produce coatings with high surface area. These approaches consist of plasma or heat spraying, thermal degradation, electro-deposition, as well as real-time activation (Gestermann & Ottaviani, 2008).

The utilization of plasma along with heat spraying methodologies necessitates the initial cleansing of a nickel base, succeeded by the application of 20 mesh alumina via sandblasting, thereby creating a surface that is conducive to optimal coating adhesion. A mixture of nickel and occasionally a secondary component like Zn, Al, Si, or just Mo within powder form with a size range of –200 to +325 mesh is introduced into a plasma gun. The plasma gun is energized at 30 kW, resulting in a temperature of approximately 5,000°C within an argon environment. The substrates are thermally regulated within a range of 200–300°C through the implementation of an exterior cooling mechanism. The blends are then applied through a spraying process, with an average rate of 25 μm each pass, as long as the required thickness is attained. Subsequently, the additional component is extracted through leaching in solutions containing sodium hydroxide in order to produce a surface area that is high. An electronic wiring spray process can be utilized as an alternative method to attain the intended surface area (Hernandez-Aldave & Andreoli, 2022).

The thermal degradation technique encompasses the application of a chloride salt mixture of the metallic material of interest, mixed with HCl or another solvent like isopropyl alcohol, onto the substrate through either spraying or just painting. This is succeeded by a high-temperature process in air, which results in the formation of the associated oxide. The aforementioned procedure bears a resemblance to the methodology employed in the creation of anodic coatings on substrates composed of titanium. Under certain circumstances, the process of reducing oxides to their metallic state can be achieved through the application of heat with the assistance of hydrogen, which may be followed by a sintering step. The metals that are amenable to application via this technique comprise Ni, Co, Ru, Zn, Rh, as well as Ir (Crook & Mousavi, 2016).

The process of electroplating, and in certain cases electroless plating, has the capability to produce elevated surface areas for a given metal or alloy. The process of plating generally entails the deposition of metallic substances or their alloys, including Ni, Mo, Co, Cd, as well as W, onto a ferrous base, which is typically composed of steel or just nickel. An alternative approach to the plating technique entails the simultaneous deposition of Ni (or various metals or metallic oxide powders) with elements like Zn or Al, followed by a chemical leaching process to produce a porous structure with a substantial surface area, if necessary. Additional alternatives comprise the application of a slender coating of precious metal, such as platinum, onto an intervening impermeable nickel layer that has been established on steel surfaces (Mendoza et al., 2017).

Electroplating is a viable method for depositing Ni–S, Ni–Zn, Ni–Re, Ni–Mo, Ni–Mo–Cd, as well as Ni–Sn coverings onto nickel surfaces. The process of preparing nickel substrates involves an initial degreasing step using trichloroethylene at ambient temperature for a duration of 30 minutes. This is followed by washing the substrates successively within acetone-ethyl alcohol as well as water. The substrates undergo activation in a concentrated solution of HNO_3 as well as glacial acetic acid in a 1:1 ratio for a duration of 5 minutes before the plating process. The deposition of Ni-S coatings can be achieved through the utilization of an improved Watts nickel bath. This bath is composed of 150 g .47.5 g .30 g I B, and 30 g KSCN, with a pH of approximately 4. The deposition process occurs at a temperature of 45°C as well as a current density of 20 mA c for a duration of 1 hour. Ni-Zn coverings were synthesized via electroplating using a solution consisting of 330 g NiS45 g 1.37.5 g B, and 20 g (pH 4.5) at a temperature of 50°C and with a current density for 3 mA cfor a duration of 1 hour. The sources cited provide information regarding the plating circumstances and baths utilized in the electroplating of Ni–Re, Ni–Mo, Ni–Med, as well as Ni–Sn (Melián-Martel et al., 2011).

The rationale behind the utilization of *in situ* activation for the purpose of attaining minimal cathodic overvoltages is motivated via the research conducted by Riley along with Moran. Their findings revealed that the introduction of minute amounts of iron into the catholyte resulted in diminished cathode possibilities on pure nickel, owing to the settling for Fe within a form with a large surface area. The augmentation in surface area was verified through the utilization of scanning electron microscopic investigations. The activation for nickel, as well as cobalt cathodes within situ, was observed to be comparable upon the introduction of Mo to a solution containing 30 wt% KOH at a temperature of 70°C. The optimal stimulation was achieved at a level of 0.004 M of Mo, which was added by means of Mo. The introduction of chloroplatinic acid into the catholyte inside chlor-alkali cells was observed to cause a reduction in cathode potentials, leading to the development of a patent. This methodology is appealing since it circumvents the expenses linked with a coating procedure. It is imperative to exercise prudence since the unprocessed chloroplatinic acid has the potential to negatively impact the final product or its intended applications, or even instigate galvanic corrosion (Domga et al., 2021).

5.4. CATHODE MATERIALS AND COMPOSITIONS

Numerous materials, as well as coatings, are being assessed for their potential utilization as cathodes within diaphragm as well as membrane cells, with a predominant focus on those that are nickel-based. Table 5.2 presents a selection of compositions disclosed within patent literature. Trasatti, along with Conway, has conducted a review regarding the kinetic factors that are involved in the HER on different materials within alkaline media. Several activation techniques have been utilized to alter the surface shape of nickel for the purpose of enhancing the surface-active region. Despite the existence of numerous potentially effective compositions and techniques, the majority of them have yet to undergo testing in chlorine cells utilized in commercial settings. The present section examines specific compositions and methods that are of particular significance while also delving into the coverings that have been employed in business practices (Millet, 2013).

5.4.1. Raney Nickel

Raney nickel was first used as a catalyst for the hydrogenation of hydrocarbons, and Justi et al. (1954) suggested it as an electrocatalyst for the HER. Several procedures are available for applying Raney nickel coatings on a given substrate (Anderson et al., 1990). These include plasma spraying, sintering of metal powders, plating from molten salt, electroplating from aqueous solutions, and dispersion plating. Ni–Al and Ni–Zn alloys are obtained by the first three methods, whereas electroplating from aqueous solutions.

Table 5.2. Catalytic Cathode Patent Literature

Patent Assignee	Elements in the Coating/ Coating Technique	Patent #
PPG Industries	Hydroxy carboxylic or aromatic acid addition to the catholyte.	U.S. 3,854,581 (1976)

Diamond Shamrock	Co-ZrO_2/flame spraying	U.S. 3,992,278 (1976)
Diamond Shamrock	Ni-Zn/electroplating	U.S. 4,104,133 (1978)
Olin Chemicals	Noble metal addition to the catholyte.	Ger. Offen. 2,818,306 (1978)
Tokuyama Soda	Ni, S/electroplating	Ger. Offen. 2,811,472 (1978)
PPG Industries	W and Ni and/or Co/electroless plating.	U.S. 4,086,149 (1979)
Osaka Soda	Ni, Mn/electroplating, thermal.	Japan 79,38,277 (1979)
Showa Denko	Ni, Al/electroplating	Japan 79,71,184 (1979)
Showa Denko	Ni, Co, P, Se/electroplating	Japan 79,75,481 (1979)
Showa Denko	Fe/oxide formation	Japan 79,80,282 (1979)
Tokuyama Soda	Ni, Ru, Ir/electroplating	Japan 79,90,080 (1979)
Chlorine Engineers	Ni, noble metals/electroplating, flame spraying.	Japan 79,110,983 (1979)
Metallgesellschaft	Fe, Co, Ni, Ag/thermal spraying	Ger. Offen. 2,829,901 (1980)
PPG Industries	Ni, P, Fe/electroplating	U.S. 4,184,941 (1980)
British Petroleum Co.	Co, Fe, Ni, Mo, W, V/thermal process	Euro Pat. Appl. 9,406 (1980)
British Petroleum Co.	Co, Fe, Ni, Mo, W, V/thermal process	Euro Pat. Appl. 9,406 (1980)
PPG Industries	Ni, Mo/flame spraying	U.S. 4,248,679 (1981)
Hooker Chemical	Ni, Zn/electroplating	U.S. 4,314,893 (1982)
Occidental Chemical	Ni, Mo, Cd/electroplating	U.S. 4,354,715 (1982)
Occidental Chemical	Cd, Pt, Pd/electroplating	U.S. 4,377,454 (1983)
Occidental Chemical	Ni, Mo, Cd/electroplating	U.S. 4,394,231 (1983)

Solvay & Cie	Co, Fe, Ru, Ni/thermal decomposition	U.S. 4,421,626 (1983)
Occidental Chemical	Ni, Mo, Cd/electroplating	U.S. 4,465,580 (1984)
Chlorine Engineers	Ni, Pt/electroplating,	U.S. 4,543,265 (1985)
Chlorine Engineers	Ni, RuO_2/electroplating, thermal decomposition	U.S. 4,555 (1986)
Solvay & Cie	Ni, Co/electroplating	U.S. 4,555 (1986)
Dow Chemical	Ru, Ni/thermal decomposition	U.S. 4,572 (1986)
ICI	Ni with Pt group metals/ electroplating	U.S. 4,586 (1986)
Asahi Kasei	Ni, Co, Cr/melt spraying	U.S. 4,605 (1986)
Oronzio de Nora	Ni, Ru oxide/electroplating	U.S. 4,618 (1986)
Oronzio de Nora	Ni, RuO_2/electroplating	U.S. 4,648 (1987)
Oronzio de Nora	Ni, Ru O_2/electroplating	U.S. 4,668 (1987)
Oronzio de Nora	Ni, RuO_2/electroplating	U.S. 4,724 (1988)
Dow Chemical	Ni, Ru oxide/thermal decomposition	U.S. 4,760 (1988)
Johnson Matthey	Pt, Ru with Au or Ag/ electroplating	U.S. 4,784 (1988)
DeNora Permelec	Ceramic/thermal deposition	U.S. 4,975 (1988)
Dow Chemical	Pd with Pt, Ni/electroplating	U.S. 5,066 (1990)
Dow Chemical	Pd with Pt, Ni/electroplating	U.S. 5,104 (1993)
Moltech	Ceramic/combination synthesis	U.S. 5,316 (1995)
Wilson Greatbatch	Ag with V/thermal decomposition	U.S. 5,389 (1996)

Source: Thomas F. O'Brien, Creative Commons License.

The application of Raney nickel to be a catalyst to facilitate the hydrogenation process of hydrocarbons dates back to its initial use, while its potential as an electromagnetic catalyst over the HER was proposed by Justi et al. in 1954. Multiple techniques exist for the deposition of Raney nickel coverings onto a specific substrate. The techniques encompassed in this category comprise plasma spraying, metallic powder sintering, melted salt plating, solution-aqueous electroplating, as well as dispersion plating. The production of Ni-Al and Ni-Zn alloys is achieved through the initial three methods, while the formation of Ni-Zn alloys is facilitated by electroplating via aqueous solutions. The deposits mentioned above undergo a treatment process involving the application of heated and settled caustic soda to facilitate the leaching of aluminum or zinc. The resulting porous layer exhibits electrocatalytic activity towards the hydrogen electrode procedure, contingent upon the initial alloy's chemical makeup and other relevant factors. The mechanical strength of the coating is typically compromised due to the leaching procedure weakening the bonding between the porous layer and the substrate. Furthermore, the subsequent processing of waste substances is a significant consideration in this methodology (Brinkmann et al., 2014; Aziz et al., 2017).

Extensive research has been conducted on the potential application of Ni-Al type Raney nickel compounds within chloralkali cells. The 3 intermediate substances of nickel as well as aluminum have been identified (Basu et al., 2013).

Hot caustic substances exhibit preferential leaching of aluminum from and . The application of a combination of compounds via the plasma-sprayed technique was discovered to exhibit activity over the hydrogen electrode's method. The most favorable outcomes were observed at a ratio of 65/35 over Ni/Al. The electrodes in question demonstrate a Tafel gradient of 40–60 mV when operating at low current densities, whereas, at high current densities, the slope increases to 120 mV (Margallo et al., 2016; Kiros & Bursell, 2008; Khasawneh et al., 2019).

The deactivation of Raney nickel-based electrodes can occur due to various factors, including the emergence of α-Ni(OH_3), which exhibits catalytic activity but lacks thermodynamic stability. The *in-situ* synthesis of *b* Ni(OH_2) has been observed to significantly enhance the long-term viability of the Raney nickel electrode. Although thermodynamically inert, this compound remains inactive. The electrochemical analysis proposes that the instability of Ni(OH)-nH_2O can be mitigated through the implementation of a cross-linked structure, as illustrated below. This approach may effectively impede the progression of oxidation as well as the consequent toxicity of the electrode's outermost layer (Van, 1998; Oloman & Radcliffe, 1986).

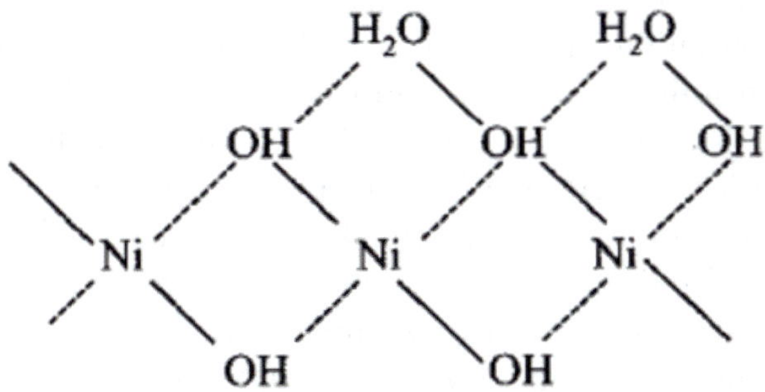

The preparation of Raney Ni is being extensively investigated by numerous researchers, including those derived from Ni–Zn alloys. These alloys can be produced via the plasma application of Ni as well as Zn powders or just electrolytic techniques. Electroplating Ni–Zn alloys leads to the creation of diverse alloys, including NiZn, $NiZn_3$, and Ni_3Zn_3. These alloys are subsequently subjected to treatment with NaOH, which serves for dissolving Zn into a soluble zincate ion. This treatment is carried out with the aim of producing a porous matrix that possesses a large surface area and demonstrates low hydrogen excessive voltage within chloralkaline cells (Park et al., 2019; Shim et al., 2015; Karlsson et al., 2016).

This chapter presents a discussion about the electrochemical assessment and production of plasma-sprayed coverings of Raney nickel as well as nickel-powdered substances, with the incorporation of various elements such as Cr, Co, Ti, W, Mo, and Fe. Additionally, the study includes electrodeposited metals of Ni-Zn as well as Co.

Remember:

The chloralkali procedure's cathode reaction generates hydrogen gas, that's frequently a valuable commodity employed in diverse industrial applications, including chemical manufacturing as well as petroleum refinery (Biddinge & Modestino, 2020).

5.4.2. Ni–Mo-Based Coatings

Ni–Mo-based coatings have been examined extensively. Inclusion of Mo in the nickel coatings to achieve catalysis was probably motivated by the use of molybdenized graphite to enhance the decomposition of sodium or potassium amalgams in the chlor-alkali industry. Although Mo is not electrocatalytically active for the HER, Ni–Mo alloys formed electrolytically or by plasma spraying show low overvoltages.

Extensive examination has been conducted on coatings based on Ni-Mo. The incorporation of Mo into nickel coatings for catalytic purposes may have been influenced by the utilization of molybdenite graphite for enhancing the breakdown of salt and potassium amalgams within the chlor-alkali sector. Despite the lack of electrocatalytic activity of Mo for the HER, the utilization of electrolytically or plasma-sprayed Ni-Mo alloys has demonstrated reduced overvoltages. It is noteworthy that bulk alloys composed of Ni-Mo are not showing reduced overvoltages. The inclusion of several minor additives, namely V, Ti, W, Cd, Re, Fe, Co, Cr, as well as Cu, in the fundamental Ni–Mo composites resulted in enhanced performance, as evidenced by the data presented in Table 5.3. The catalytic performance of the cathode composed of Ni-Mo alloy exhibits significant variability depending on its chemical composition as well as the preparation technique employed. The alloy, composed of 60 atomic percent nickel and 40 atomic percent molybdenum, exhibited a notable level of catalytic efficacy (Farzami et al., 2011; Garcia-Herrero et al., 2018).

Table 5.3. Characteristics of the Ni–Mo-based Alloy

	Tafel Slope		
Material	**At <50 (mA c)**	**At >50 (mA)**	**Overvoltage at 300 (mA)**
Ni–Mo–Fe	115	165	187
Ni–Mo–Cu	28	180	190
Ni–Mo–Zn	24	116	220
Ni–Mo–W	28	150	286
Ni–Mao	26	140	270
Ni–Mo–Cr	32	165	350
Steel	135	125	540

Source: Tilak V. Bommaraju, Creative Commons License.

The Ni-Med covering, formed electrolytically upon nickel substrates and studied extensively from an essential perspective, is a noteworthy, patented structure. It displayed a Tafel slope of approximately 40 mV within the current density spectrum of 0.1–0.3 A c. The research findings indicate that the coatings exhibit higher activity levels compared to Ni-Mo layers. This enhanced activity is explained by the inclusion of Cd, which is responsible for conferring the capacity for absorption of hydrogen or modulating the characteristics associated with the hydrogen stage upon the surface. The present study describes coatings that depend on Ni-Mo-Fe (Joudaki et al., 2009; Burney, 1993).

5.4.3. Other Ni-Based Materials

The Ni-Sn metal, having an Sn concentration ranging from 5.9 to 37 at%, was synthesized through electroplating using a bath solution comprising Ni, pyrophosphate, as well as Sn. The resultant alloy exhibited a low hydrogen excessive voltage within 28% KOH at 20°C, using an average Tafel slope of 52 mV. Conversely, the HER did not display reduced overvoltage in applied Ni-Mn alloys. It is noteworthy that coatings constructed from Ni-Si exhibited minimal overvoltage as well as Tafel slopes. The observed behavior of the electrodes can be attributed to the significant increase in surface area resulting from the leaching of silicon with sodium hydroxide (Franco et al., 2019; Zheng et al., 2011).

The electroless coating method was employed to prepare Ni–Co–P, Ni–P, and Co-P, and the resulting hydrogen overvoltage proved to be comparatively low in KOH solutions, with a Tafel slope of 120 mV. The observed trend indicates that the hydrogen overvoltage is positively correlated with the P material of the covering.

The investigated compositions encompass large-area cobalt coverings, amorphous alloys containing Ni, Co, Si, as well as B, composite materials of Ni + Al+ , PTFE-bonded metallic powders, nickel coverings adapted with hetero polymer acids, ABs-type cathodes such as La, deposited by electrolysis Fe, Ni, Mo coats co-formed using W or just W/Mo, spinel's composed of Ni, Co, Fe, Mn, as well as La, as well as Zr, Al, and (Otashu & Baldea, 2019; Schneiders et al., 2001).

5.4.4. Oxides

The oxides for noble metals have been studied as cathode material within a distinct category of substances. The reduction of these oxides is thermodynamically favorable. The stability of cathodes for the HER is determined by the reduction kinetics. The stability of was observed to persist over extended periods of electrolysis when utilized as a cathode without a significant decrease of the oxide. It should be noted, however, that RuO_2

can undergo reduction from hydrogen at elevated temperatures (Roh et al., 2019).

Several noble metal coverings, such as , were analyzed within alkaline solutions. The minimal hydrogen overvoltage exhibited by these materials can be attributed to their substantial surface area, that's a result of their structure as well as the preparation technique employed. The physicochemical characteristics of Ir, which was produced through the thermal oxidization for at levels below 350°C, exhibited variations compared to those synthesized at greater temperatures. The ideal temperature for achieving complete oxidation of was approximately 500°C.

The Tafel slope of the Hydrogen Electrode response on a heat formed cathode exhibits a range of 40–50 mV at minimal current concentrations as well as 120 mV that high rates of current. The coated cathodes facilitated the cathodic elimination of within chlorate solutions.

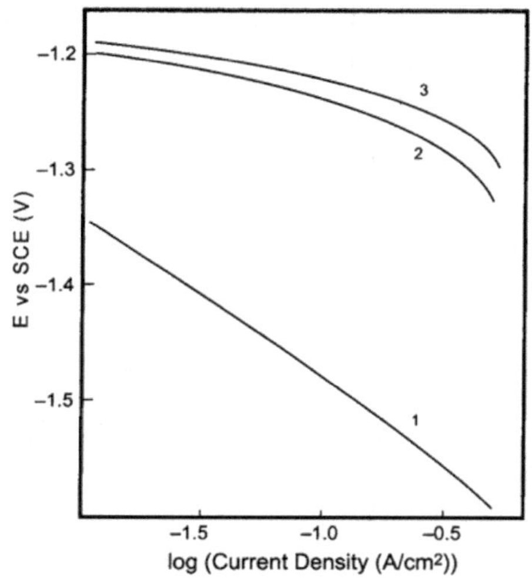

Figure 5.9. *Illustration of polarization data for the HER in 35 wt.% NaOH at 85°C on (i) electroless Ni coating; (ii) thermally formed; and (iii) with electroless Ni coating.*

Source: S. Trasati, Creative Commons License.

The present study reports the successful fabrication of a composite permeable layer comprising Ni and , exhibiting a remarkable catalytic activity towards the transformation of hydrogen. The fabrication involved the dispersion electroplated technique, wherein a Watts-type bath that included powder was utilized. The study reports that a nickel substrate coated with heat formed , as well as altered by an additional layer regarding electroless nickel, exhibited an operational potential of 250 mV compared to an even nickel electrode (as shown in Figure 5.9) for a period exceeding 2.5 years. Additionally, the coated substrate demonstrated a relatively low sensitivity to the existence of iron (Pletcher, 1984).

5.4.5. Sulfides

Cathodes activated by sulfur were initially devised in the latter part of the 1970s for the purpose of water electrolysis as well as subsequently expanded for application in chlor-alkali cells. The compounds , , and , have undergone examination. However, has been subjected to the most extensive study (Karakilcik et al., 2019; Dötzel & Schneider, 2002).

, can be synthesized through several methods, including electrodeposition using a Watts bath that contains sulfur compounds like thiourea or just thiocyanates, electroplated within a solution of dithionite, distribution plating using a Watts bath alongside powders, and sulfidation of NiO. The level of thiourea by the solution used to plat directly correlates with the amount of sulfur present in the deposit. As the amount of sulfur increases, there is a decrease within the hydrogen overvoltage, with the optimal level being 12%.

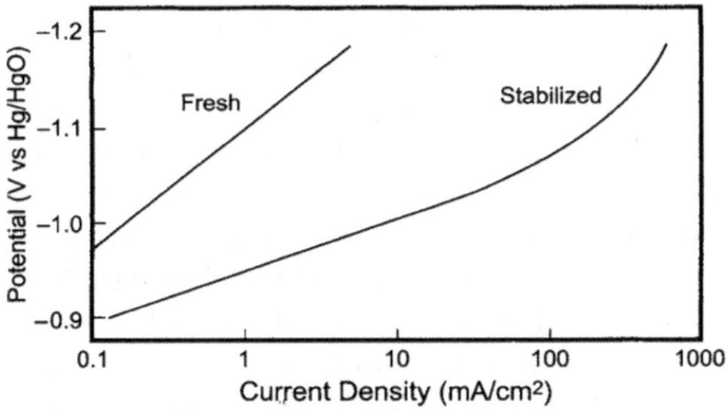

Figure 5.10. *Illustration of polarization data for the HER in IM NaOH at 25°C on Ni electrode.*

Source: S. Trasati, Creative Commons License.

The polarization arcs generated by the cathode within alkaline solutions exhibit two distinct Tafel regions with varying slopes, dependent on the manufacturing technique and the concentration of sulfur present. These observations are indicative of a steady-state system. The polarized features of the cathode are depicted in Figure 5.10. The initial cathode overpotential exhibited a considerable elevation, with a Tafel gradient of approximately 120 millivolts. The voltage surge experienced a gradual decline over the course of the electrolysis process, ultimately reaching a state of stabilization after approximately 200 hours. At minimal current density levels, the Tafel slope exhibited a decrease to 60 mV, whereas, at high levels of current, it remained at a high level (Sugiyama et al., 2003; Kaveh et al., 2009).

The cathodes of underwent irreversible deactivation when exposed to , , and due to the hindrance caused by these contaminants in the adhesion of H intermediaries upon the cathode's outer layer.

The NiSg layer that underwent electrodeposition was subject to modification through the incorporation of additional elements aimed at mitigating the processes of degradation as well as deactivation. The introduction of molybdenum enhanced permeability to impurities, particularly iron, due to the significant increase in surface area. Additional sulfur-based formulations that have been investigated include activated and Teflon-bonded formulations. The chemical compound denoted as Niis of interest in materials science (Busto et al., 2011; Špalek, 1994).

5.4.6. In Situ Activation

The diverse cathode coatings that were analyzed for implementation in manufacturing chlor-alkali cells, as previously discussed, necessitate electroplating or just thermal techniques for applying them onto a nickel or simply steel cathode base. The cost of the process in question is dependent on the method utilized for the application of the active coating. However, it is possible to circumvent this expense by depositing an electro-catalytically active substance *in situ*. This process is achieved by introducing a precursor into either the catholic solution of the membrane cells or just the anolyte of the diaphragm cells. The precursor will then undergo reduction to its metallic state according to cathodic polarization circumstances. Among the various additives considered, chloroplatinic acid appears to be a promising option, according to a cost-benefit evaluation. The addition of a small amount of within the catholyte might end up in a reduction of voltage by 100–150 mV. This effect can persist for 2–3 months, contingent upon the composition and quantity of impurities within the catholyte flow, particularly iron (Trasatti, 1992; Bergner, 1982; Grotheer et al., 2006).

5.5. HER MECHANISMS ON LOW OVERVOLTAGE CATHODES

Section 5.2 delineates the reaction pathways implicated in the progression of the HER on flat metallic surfaces. The HER system in alkaline substances is commonly understood as involving the utilization of water molecules to be a proton source. These protons are subsequently discharged onto the cathode's outermost layer, M, resulting in the formation of an M–H bond. The hydrogen that has been adsorbed has the potential to either undergo a combination reaction with a different hydrogen molecule, resulting in the formation of (as shown in Eqn. (3)), or alternatively, it could interact using a water molecule and create .

$$H_2O + M + e \rightarrow MH + OH^-$$

$$2MH \rightarrow H_2 + 2M$$

$$MH + H_2O + e \rightarrow H_2 + M + OH^-$$

The Tafel slopes exhibit reduced magnitudes when adsorbed intermediaries are implicated in the rate-determining step of the HER, as evidenced by theoretical predictions and illustrations. The occurrence of low Tafel gradients can be attributed to processes that are contingent upon the potential-dependent coverage for adsorbed intermediaries. However, it is worth noting that such mechanisms may pose a drawback on metals due to the possibility of embrittlement caused by hydrogen arising from the spreading of retained hydrogen through the metal. Although low Tafel gradients offer voltage benefits, they can significantly impact long-term efficiency in a negative manner (Garcia-Herrero et al., 2017; Jalali, 2009).

The findings indicate that an increase within surface area leads to a decrease in the Tafel slope when comparing polarization measurements from massive surface area planning to comparable metal in mass form. The Tafel slope exhibited by uniform Ni in alkaline fluids experiences a reduction from approximately 100 mV to 30 mV upon the utilization of Raney Ni, along with assorted compositions, as depicted in Figure 5.11. The aforementioned conduct is not exclusive to Nickel (Ni), as it is also witnessed in Platinum (Pt) throughout the progression of the HER within alkaline environments (Figure 5.12).

As the HER occurred on high-surface area matrices, the effect of the porous structure on the Tafel slope behavior was examined by assuming a homogeneous elementary flooded pore model and deducing the current–

voltage relationships, taking into account the backward rate and the coverage dependence on potential. This analysis showed that when the rate-determining step is discharge or electrochemical desorption, the Tafel slopes change from 2AF/ F and 2RT/3 F,respectively to 4RT/ F at high over voltages.

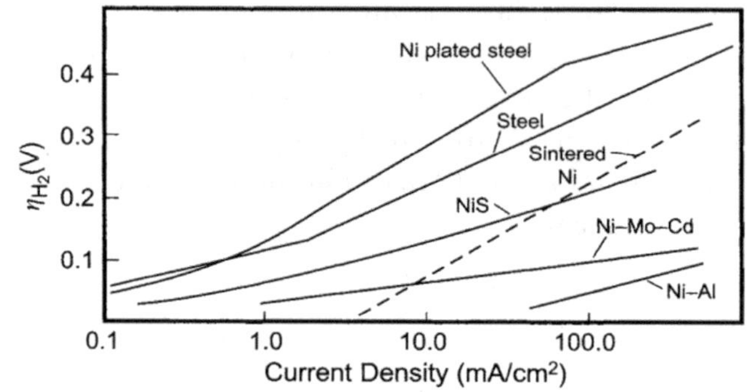

Figure 5.11. *Illustration of polarization data for the HER on various Ni-based compositions in 15% NaOH +17% NaCl solutions at 95°C.*

Source: C. P. Chen, Creative Commons License.

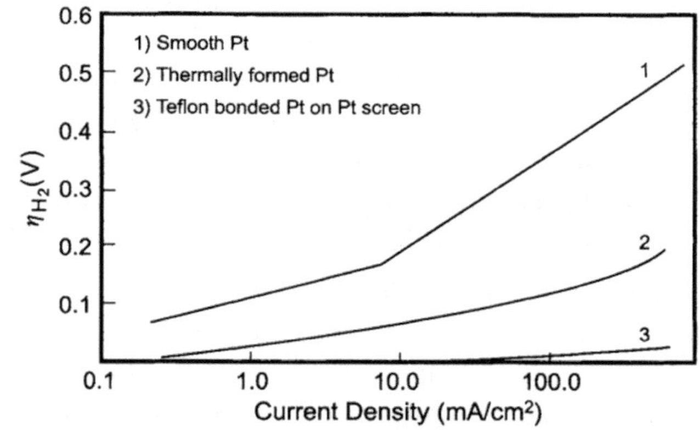

Figure 5.12. *Illustration of polarization data for the HER on smooth and high surface area Pt in 15% NaOH +17% NaCl solutions at 95°C.*

Source: C. P. Chen, Creative Commons License.

The impact of a porous framework on the incline of Ta conduct was investigated in light of the occurrence of the HER upon high-surface-area matrix structures. This was achieved by adopting a homogeneous fundamental flooded pores model and inferring the current-voltage associations while considering the potential coverage reliance along with the reversed rate. The present study demonstrates that in cases where the discharge or just electrochemical desorption serves as the rate-determining phase, the gradients of Tafel undergo a transition from 2AF/F and 2RT/3 F into 4RT/F at elevated overvoltage. In cases where recombination seems to be the rate-limiting step, the gradient of Tafel undergoes a transition from RT/2F into a non-Tafel behavior, where the current $I\infty$. Hence, the minimal Tafel gradients observed via high-surface composites cannot be attributed to the porous texture of the outermost structure (Cameron et al., 1990; Furuya & Aikawa, 1999).

An alternative interpretation, predicated on an evaluation of the decline in electrical potential subsequent to the cessation of current, posits the potential emergence of a hydride lattice with three dimensions (on high surface area nickel substrates produced through the dissolution of plasma-sprayed Ni–Al coverings along with electrodeposited Ni–Med coatings), which facilitates the HER, as explicated below.

$$x\mathrm{Ni} + \mathrm{H_2O} + \mathrm{e} \rightleftharpoons \mathrm{Ni}_x\mathrm{H} + \mathrm{OH}^-$$

$$\mathrm{Ni}_x\mathrm{H} + \mathrm{H_2O} + \mathrm{e} \rightleftharpoons (\mathrm{Ni}_x\mathrm{H})\mathrm{H}_{ads} + \mathrm{OH}^-$$

$$(\mathrm{Ni}_x\mathrm{H})\mathrm{H}_{ads} + \mathrm{H_2O} + \mathrm{e} \rightarrow \mathrm{Ni}_x\mathrm{H} + \mathrm{OH}^- + \mathrm{H_2}$$

Although the low gradient of Tafel can be achieved by the two distinct quasi-equilibria that precedes the slow step, it remains challenging to understand why hydride formation occurs solely on large surface area matrix structures rather than smooth electrodes. Moreover, as previously mentioned, this conduct is not exclusive to nickel, as comparable conduct was witnessed on Pt electrodes (*refer* Figure 5.12). The active sites of various energetics within high-surface electrodes may lead to alterations in activation energy, thereby causing mechanistic changes. Nevertheless, the precise understanding of the influence of surface area upon the induction of mechanistic variations remains to be fully elucidated. It is imperative that any model put forth incorporates a factor associated with a surface region within the exponential form β.

The Rubased electrodes are thought to exhibit low Tafel gradients of 40–50 mV due to the process of the formation of hydrogen:

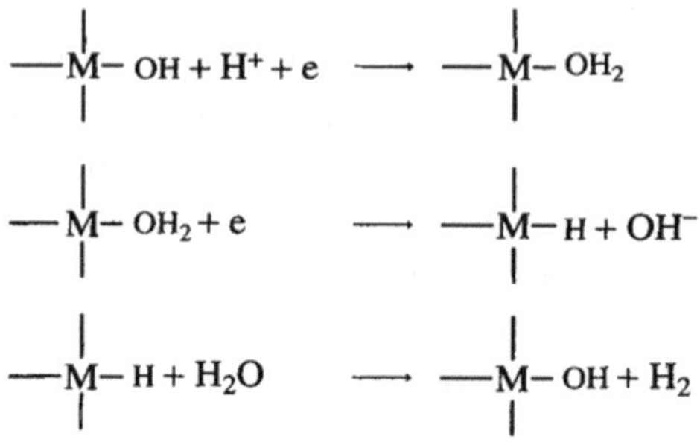

This mechanism enables the preservation of the oxide state without any alteration.

5.6. CATHODE DEACTIVATION

There exist various factors that may contribute to the augmentation of cathode potentials. Some factors that can cause blockage are the presence of gas bubbles and organic contaminants. In diaphragmatic cells, inorganic contaminants like silicates and deposition of substantial overvoltage metallic substances (e.g., Hg, Cd) upon the cathode base, as well as corrosion within the catalytic coverings, are common issues. These issues may arise during shutdowns due to reverse currents, a target by hypochlorite, as well as the development of hydride, which may lead to spalling of the coverings or the base material (Stefan et al., 2010).

Research has indicated that the cathode capacity for steel is negatively impacted by the presence of Cr, Ni, Hg, Fe, as well as organic substances containing N or just S. The study revealed that activated cathodes experienced a significant decline in performance due to the presence of iron, as depicted in Figure 5.13. Cathodes coated with Pt-Ru are susceptible to the influence of iron (Fe) even at concentrations as low as 3 parts per million (ppm) within the solution. Nonetheless, it has been observed that in the presence of smooth cathodes made of Ni or Ti, the cathode potential is reduced by Fe. This is attributed to the fact that the deposits developed within the cathode possess a considerable surface area.

Cathode deactivation can also occur due to the gradual dissolution of a constituent element within the cathode coatings. The deactivation of a cathode occurs gradually due to the disintegration of S within alkaline solutions. It is plausible to expect analogous conduct in the event that any of the constituents present in the covering exhibit thermodynamic instability when subjected to a NaOH solution, like in the case of Mo within Ni–Mo coverings. Therefore, it is crucial to evaluate the preliminary thermodynamic resilience of the substrate along with coating elements using the Pourbaix schematics. This analysis has the potential to offer valuable insights into the possible mechanisms that could lead to an upsurge in overvoltage (Lim et al., 2023; Cole, 1982).

Both the cathode covering, and substrate must exhibit resistance to hypochlorite as well as reverse currents that are typically encountered throughout shutdowns. The interruption of current results in an opposite current within the cell, which can be attributed to the engaged chlorine species within the anolyte. Consequently, there is a transient shift in the cathode's role to that of an anode until the entirety of the engaged chlorine species undergoes reduction to form chloride. Subsequent to a cessation of operations, the hypochlorite present in the anolyte is conveyed to the catholyte, thereby elevating its hypochlorite concentration to a range of 5 to 30 parts per million in membrane cells as well as 300 to 5,000 parts per million in diaphragmatic cells.

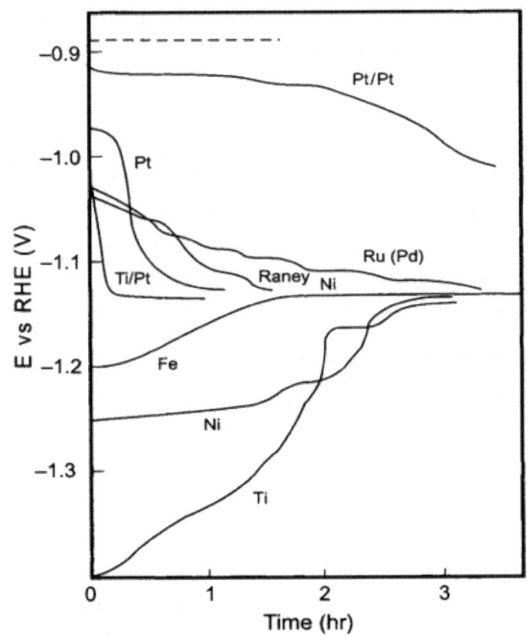

Figure 5.13. *Variation of the cathode potential with time during the course of the HER from 33% NaOH at 90°C and 3 kA in the presence of 10 ppm Fe. The dashed line represents the reversible potential.*

Source: Thomas F. O'Brien, Creative Commons License.

$$OCl^- + H_2O + 2e \rightarrow Cl^- + 2OH^- \quad \text{(cathodic reaction)}$$

$$M \rightarrow M^{2+} + 2e \quad \text{(anodic reaction)}$$

The removal of cathode coverings is primarily attributed to contaminants and within the catholyte. Coverings that exhibit high porosity and a substantial surface area have the potential to mitigate the deleterious effects of impurity poisoning. This is due to the fact that the electrode provides a considerable area for the HER, even when a portion of its outermost layer becomes inaccessible. The shape and structure of the active layer can determine whether porosity functions as a kind of sponge over contaminants. It is important to acknowledge that there exists no catalyzing cathode coating that is impervious to the influence of impurities, as these impurities have the potential to either deposit or obstruct the surface on a physical level. In the event of an elevated hydrogen excessive voltage affecting the deposited creatures, there is a possibility of an increase in overvoltage, which could nullify the intended outcomes. In the event that the cathode coverings comprise oxides or just sulfides, it's plausible that the contaminants may exhibit non-deposition tendencies in substrates that are considered "nonmetallic."

The deleterious consequences arising from the presence of dissolved chlorine within the anolyte might be mitigated through two methods: the eradication of hypochlorite upon cessation of current or just the implementation of cathodic protection for the cathode throughout shutdowns Hypochlorite removal can be achieved by flushing the anolyte or by reacting it with a sulfite or H2Oa the cathode material is unstable in the pH region of 3–7, the pH must simultaneously be brought into the alkaline region. This step must be implemented as quickly as possible (Carvela et al., 2020; Reig et al., 2014).

5.7. SUMMARY

Cathodes are of paramount importance during the electrolysis procedure employed by the chlor-alkali sector to generate chlorine, caustic soda, as well as hydrogen gases. Electrolysis occurs within an electrolytic cell, which may be referred to as a diaphragmatic cell, mercury cell, or just membrane cell, according to the particular technology utilized (Mihaiescu et al., 2012). The diaphragmatic cell process involves the implementation of a diaphragm composed of either asbestos or just polymer to segregate the anode as well as the cathode sections. The mercury-based cell process involves utilizing an insoluble mercury pool to be the cathode. In recent times, the utilization of membrane cell technologies has garnered significant attention owing to the environmental apprehensions linked with asbestos as well as mercury. Within electrolytic devices, cathodes play a crucial role in enabling the electrolysis for the brine to produce chlorine, caustic soda, as well as hydrogen gas.

REFERENCES

1. Anderson, E., Taylor, E. J., Vilambi, N. R. K., & Gelb, A., (1990). Preliminary analysis of fuel cell derived technologies applied to energy conservation in the chlor-alkali industry. *Separation Science and Technology, 25*(13–15), 1537–1554.
2. Antozzi, A. L., Bargioni, C., Iacopetti, L., Musiani, M., & Vázquez-Gómez, L., (2008). EIS study of the service life of activated cathodes for the hydrogen evolution reaction in the chlor–alkali membrane cell process. *Electrochimica Acta, 53*(25), 7410–7416.
3. Aziz, H. A., Ghazali, M. F., Yusoff, M. S., & Hung, Y. T., (2017). Waste treatment and management in chlor-alkali industries. In: *Handbook of Advanced Industrial and Hazardous Wastes Management* (pp. 611–655). CRC Press.
4. Bacquart, T., Murugan, A., Carré, M., Gozlan, B., Auprêtre, F., Haloua, F., & Aarhaug, T. A., (2018). Probability of occurrence of ISO 14687–2 contaminants in hydrogen: Principles and examples from steam methane reforming and electrolysis (water and chlor-alkali) production processes model. *International Journal of Hydrogen Energy, 43*(26), 11872–11883.
5. Basu, S., Mukhopadhyay, S. K., Gangopadhyay, A., & Dastidar, S. G., (2013). Characteristic change of effluent from a chlor-alkali industry of India due to process modification. *International Research Journal of Environment Sciences, 2*(2), 44–47.
6. Bergner, D., (1982). Membrane cells for chlor-alkali electrolysis. *Journal of Applied Electrochemistry, 12*(6), 631–644.
7. Biddinger, E. J., & Modestino, M. A., (2020). Electro-organic syntheses for green chemical manufacturing. *The Electrochemical Society Interface, 29*(3), 43.
8. Brinkmann, T., Santonja, G. G., Schorcht, F., Roudier, S., & Sancho, L. D., (2014). *Best Available Techniques (BAT) Reference Document for the Production of Chlor-Alkali*. JRC (Joint Research Center EU), JRC91156.
9. Burney, H. S., (1993). Membrane chlor-alkali process. *Modern Aspects of Electrochemistry*, *24*, 393–438.
10. Busto, Y., Cabrera, X., Tack, F. M. G., & Verloo, M. G., (2011). Potential of thermal treatment for decontamination of mercury containing wastes from chlor-alkali industry. *Journal of Hazardous Materials, 186*(1), 114–118.
11. Cameron, D. S., Phillips, R. L., & Willis, P. M., (1990). Poison tolerant platinum catalyzed cathodes for membrane cells. *Modern Chlor-Alkali*

Technology, 4, 95–107.

12. Carvela, M., Conti, J., Lobato, J., Scialdone, O., & Rodrigo, M. A., (2020). Effect of the anode composition on the performance of reversible chlor-alkali electro-absorption cells. *Separation and Purification Technology, 248*, 117017.
13. Chatenet, M., Aurousseau, M., Durand, R., & Andolfatto, F., (2003). Silver-platinum bimetallic catalysts for oxygen cathodes in chlor-alkali electrolysis: Comparison with pure platinum. *Journal of the Electrochemical Society, 150*(3), D47.
14. Chatterjee, S. N., (1984). Design and development of electrochemical chlor-alkali cells. *Electrochemical Cell Design*, 1–24.
15. Chavan, N., Pinnow, S., Polcyn, G. D., & Turek, T., (2015). Non-isothermal model for an industrial chlor-alkali cell with oxygen-depolarized cathode. *Journal of Applied Electrochemistry, 45*, 899–912.
16. Cole, J. C., (1982). Future trends in the chlor-alkali industry. *Electrochemistry in Industry: New Directions*, 355–359.
17. Crook, J., & Mousavi, A., (2016). The chlor-alkali process: A review of history and pollution. *Environmental Forensics, 17*(3), 211–217.
18. Domga, T., Noumi, G. B., Sieliechi, M. J., & Tchatchueng, J. B., (2021). Synthesis of nitrogen and phosphorus co-doped graphene as efficient electrocatalyst for oxygen reduction reaction under strong alkaline media in advanced chlor-alkali cell. *Carbon Trends, 4*, 100043.
19. Dötzel, O., & Schneider, L., (2002). Non-asbestos diaphragms in chlor-alkali electrolysis. *Chemical Engineering and Technology, 25*(2), 167–171.
20. Endoh, E., Otouma, H., & Morimoto, T., (1988). Advanced low hydrogen overvoltage cathode for chlor-alkali electrolysis cells. *International Journal of Hydrogen Energy, 13*(4), 207–213.
21. Farzami, F., Joudaki, E., & Hashemi, S. J., (2011). Comparative study on application of bimetallic Pt-based alloy electrocatalysts in advanced chlor-alkali electrolysis. *Engineering, 3*(8), 836.
22. Figueiredo, R. S., Bertazzoli, R., & Rodrigues, C. A., (2013). Copper/carbon/PTFE oxygen-depolarized cathodes for chlor-alkali membrane cells. *Industrial and Engineering Chemistry Research, 52*(16), 5611–5615.
23. Franco, F., Prior, J., Velizarov, S., & Mendes, A., (2019). A systematic performance history analysis of a chlor-alkali membrane electrolyzer under industrial operating conditions. *Applied Sciences, 9*(2), 284.
24. Furuya, N., & Aikawa, H., (1999). A study of the gas diffusion electrodes for

chlor-alkali membrane cell. In: *The Electrochemical Society Proceedings* (pp. 180–187).

25. Furuya, N., & Aikawa, H., (2000). Comparative study of oxygen cathodes loaded with Ag and Pt catalysts in chlor-alkali membrane cells. *Electrochimica Acta, 45*(25–26), 4251–4256.
26. Garcia-Herrero, I., Margallo, M., Laso, J., Onandía, R., Irabien, A., & Aldaco, R., (2017). Measuring the vulnerability of an energy intensive sector to the EU ETS under a life cycle approach: The case of the chlor-alkali industry. *Sustainability, 9*(5), 837.
27. Garcia-Herrero, I., Margallo, M., Onandía, R., Aldaco, R., & Irabien, A., (2017). Life cycle assessment model for the chlor-alkali process: A comprehensive review of resources and available technologies. *Sustainable Production and Consumption, 12*, 44–58.
28. Garcia-Herrero, I., Margallo, M., Onandía, R., Aldaco, R., & Irabien, A., (2017). Environmental challenges of the chlor-alkali production: Seeking answers from a life cycle approach. *Science of the Total Environment, 580*, 147–157.
29. Garcia-Herrero, I., Margallo, M., Onandía, R., Aldaco, R., & Irabien, A., (2018). Connecting wastes to resources for clean technologies in the chlor-alkali industry: a life cycle approach. *Clean Technologies and Environmental Policy, 20*, 229–242.
30. Gebhard, M., Franzen, D., Paulisch, M., Tichter, T., Turek, T., Manke, I., & Roth, C., (2020). High performance oxygen depolarized cathodes for chlor-alkali electrolysis using electrospun poly(vinylidene fluoride) (PVDF) barrier layers. In: *Electrochemical Society Meeting Abstracts 237* (No. 21, pp. 1276–1276). The Electrochemical Society, Inc.
31. Gebhard, M., Paulisch, M., Hilger, A., Franzen, D., Ellendorff, B., Turek, T., & Roth, C., (2019). Design of an in-operando cell for X-ray and neutron imaging of oxygen-depolarized cathodes in chlor-alkali electrolysis. *Materials, 12*(8), 1275.
32. Gestermann, F., & Ottaviani, A., (2008). Chlorine production with oxygen-depolarized cathodes on an industrial scale. In: John, M., (ed.), *Modern Chlor-Alkali Technology* (Vol. 8, pp. 49–56).
33. Ghoshal, S., Li, J., & Mukerjee, S., (2015). Non-noble metal catalysts as oxygen depolarized cathodes in chlor alkali electrolyzers. In: *Electrochemical Society Meeting Abstracts 227* (No. 16, pp. 1236–1236). The Electrochemical Society, Inc.
34. Gilliam, R. J., Boggs, B. K., Decker, V., Kostowskyj, M. A., Gorer, S., Albrecht, T. A., & Bard, A. J., (2012). Low voltage electrochemical process

for direct carbon dioxide sequestration. *Journal of the Electrochemical Society, 159*(5), B627.

35. Giz, M. J. D., Marengo, M. C., Ticianelli, E. A., & Gonzalez, E. R., (2003). Electrochemical and physical characterization of Ni-Cu-Fe alloy for chlor-alkali hydrogen cathodes. *Ecletica Quimica, 28*, 21–28.
36. Grotheer, M., Alkire, R. C., Varjian, R., Srinivasan, V., & Weidner, J. W., (2006). Industrial electrolysis and electrochemical engineering. *The Electrochemical Society Interface, 15*(1), 52.
37. Guo, J. H., & Sun, W. Y., (2020). Integrating nickel-nitrogen doped carbon catalyzed CO_2 electroreduction with chlor-alkali process for CO, Cl_2, and $KHCO_3$ production with enhanced techno-economics. *Applied Catalysis B: Environmental, 275*, 119154.
38. Hachiya, T., Sasaki, T., Tsuchida, K., & Houda, H., (2009). Ruthenium oxide cathodes for chlor-alkali electrolysis. *ECS Transactions, 16*(39), 31.
39. Hernandez-Aldave, S., & Andreoli, E., (2022). Oxygen depolarized cathode as a learning platform for CO 2 gas diffusion electrodes. *Catalysis Science and Technology*.
40. Hine, F., Tilak, B. V., & Viswanathan, K., (1986). Chemistry and chemical engineering in the chlor-alkali industry. *Modern Aspects of Electrochemistry, 18*, 249–302.
41. Hou, M., Chen, L., Guo, Z., Dong, X., Wang, Y., & Xia, Y., (2018). A clean and membrane-free chlor-alkali process with decoupled Cl_2 and H_2/NaOH production. *Nature Communications, 9*(1), 438.
42. Ichinose, O., Kawaguchi, M., & Furuya, N., (2004). Effect of silver catalyst on the activity and mechanism of a gas diffusion type oxygen cathode for chlor-alkali electrolysis. *Journal of Applied Electrochemistry, 34*(1).
43. Ito, H., & Manabe, A., (2022). Chlor–alkali electrolysis. In: *Electrochemical Power Sources: Fundamentals, Systems, and Applications* (pp. 281–304). Elsevier.
44. Iwakura, C., Tanaka, M., Nakamatsu, S., Inoue, H., Matsuoka, M., & Furukawa, N., (1995). Electrochemical properties of Ni (Ni+ RuO_2) active cathodes for hydrogen evolution in chlor-alkali electrolysis. *Electrochimica Acta, 40*(8), 977–982.
45. Jaccaud, M., Leroux, F., & Millet, J. C., (1989). New chlor-alkali activated cathodes. *Materials Chemistry and Physics, 22*(1–2), 105–119.
46. Jalali, A. A., Mohammadi, F., & Ashrafizadeh, S. N., (2009). Effects of process conditions on cell voltage, current efficiency, and voltage balance

of a chlor-alkali membrane cell. *Desalination, 237*(1–3), 126–139.

47. Jing, W., & Chuanwei, Z., (2011). Electrochemical properties of Fe-Ni-Zr active cathodes processed by VIM technique for hydrogen evolution in chlor-alkali electrolysis. In: *2011 International Conference on Materials for Renewable Energy and Environment* (Vol. 1, pp. 721–724). IEEE.
48. Jorne, J., & Louvar, J. F., (1980). Gas-diverting electrodes in the chlor-alkali membrane cell. *Journal of the Electrochemical Society, 127*(2), 298.
49. Joudaki, E., Farzami, F., Mahdavi, V., & Hashemi, S. J., (2010). Performance evaluation of oxygen-depolarized cathode with PtPd/C electrocatalyst layer in advanced chlor-alkali cell. *Chemical Engineering and Technology, 33*(9), 1525–1530.
50. Joudaki, E., Hashemi, S. J., Mohammadi, F., Yousefi, A., & Eivazkhani, M., (2011). Oxygen reduction electrode in advanced chlor-alkali with ruthenium as electro-catalyst. *The Canadian Journal of Chemical Engineering, 89*(1), 197–201.
51. Joudaki, E., Mohammadi, F., Yousefi, A., Eivazkhani, M., & Yousefi, M., (2009). Comparative study of the Pt and Ru carbon support's affection on the caustic current efficiency in a zero-gap advanced chlor-alkali cell. *Aust. J. Basic Appl. Sci., 3*(3), 1879–1887.
52. Jović, B. M., Lačnjevac, U. Č., Krstajić, N. V., & Jović, V. D., (2014). Service life test of the NiSn coatings as cathodes for hydrogen evolution in industrial chlor-alkali electrolysis. *International Journal of Hydrogen Energy, 39*(17), 8947–8958.
53. Jung, J., Postels, S., & Bardow, A., (2014). Cleaner chlorine production using oxygen depolarized cathodes? A life cycle assessment. *Journal of Cleaner Production, 80*, 46–56.
54. Karakilcik, H., Erden, M., & Karakilcik, M., (2019). Investigation of hydrogen production performance of chlor-alkali cell integrated into a power generation system based on geothermal resources. *International Journal of Hydrogen Energy, 44*(27), 14145–14150.
55. Karlsson, R. K., & Cornell, A., (2016). Selectivity between oxygen and chlorine evolution in the chlor-alkali and chlorate processes. *Chemical Reviews, 116*(5), 2982–3028.
56. Kaveh, N. S., Mohammadi, F., & Ashrafizadeh, S. N., (2009). Prediction of cell voltage and current efficiency in a lab scale chlor-alkali membrane cell based on support vector machines. *Chemical Engineering Journal, 147*(2–3), 161–172.
57. Kelham, S. F., (1991). Chlor-alkali products. *An Introduction to Industrial Chemistry*, pp. 284–303.

58. Khasawneh, H., Saidan, M. N., & Al-Addous, M., (2019). Utilization of hydrogen as clean energy resource in chlor-alkali process. *Energy Exploration and Exploitation, 37*(3), 1053–1072.
59. Kiros, Y., & Bursell, M., (2008). Low energy consumption in chlor-alkali cells using oxygen reduction electrodes. *Int. J. Electrochem. Sci., 3*, 444–451.
60. Kiros, Y., Pirjamali, M., & Bursell, M., (2006). Oxygen reduction electrodes for electrolysis in chlor-alkali cells. *Electrochimica Acta, 51*(16), 3346–3350.
61. Lakshmanan, S., & Murugesan, T., (2014). The chlor-alkali process: Work in progress. *Clean Technologies and Environmental Policy, 16*, 225–234.
62. Leddy, J. J., (1980). The chlor-alkali industry. *Journal of Chemical Education, 57*(9), 640.
63. Li, K., Fan, Q., Chuai, H., Liu, H., Zhang, S., & Ma, X., (2021). Revisiting chlor-alkali electrolyzers: From materials to devices. *Transactions of Tianjin University, 27*, 202–216.
64. Lim, H. W., Park, J. H., Yan, B., Kim, J. Y., & Lee, C. W., (2023). Liquid-diffusion electrode with core-shell structured mixed metal oxide catalyst for near-zero polarization in chlor-alkali electrolysis. *Applied Catalysis B: Environmental, 322*, 122095.
65. Lipp, L., Gottesfeld, S., & Chlistunoff, J., (2005). Peroxide formation in a zero-gap chlor-alkali cell with an oxygen-depolarized cathode. *Journal of Applied Electrochemistry, 35*, 1015–1024.
66. Margallo, M., Onandía, R., Aldaco, R., & Irabien, A., (2016). When life cycle thinking is necessary for decision making: Emerging cleaner technologies in the chlor-alkali industry. *Chemical Engineering Transactions, 52*, 475–480.
67. Melián-Martel, N., Sadhwani, J. J., & Báez, S. O. P., (2011). Saline waste disposal reuse for desalination plants for the chlor-alkali industry: The particular case of Pozo Izquierdo SWRO desalination plant. *Desalination, 281*, 35–41.
68. Mendoza, P. G., Moshtarikhah, S., Langenhan, A. S., de Groot, M. T., Keurentjes, J. T. F., Schouten, J. C., & van der Schaaf, J., (2017). Intensification of the chlor-alkali process by using a spinning disc membrane electrolyzer. *Chemical Engineering Research and Design, 128*, 120–129.
69. Mihaiescu, T., Mihaiescu, R., & Odagiu, A., (2012). Environmental issues within the Chlor-alkali manufacturing industry-mercury cell process. *Bull. Univ. Agric. Sci. Vet. Med., 69*, 168–176.

70. Millet, P., (2013). Chlor-alkali technology: Fundamentals, processes, and materials for diaphragms and membranes. In: *Handbook of Membrane Reactors* (pp. 384–415). Woodhead Publishing.
71. Mirzazadeh, T., Mohammadi, F., Soltanieh, M., & Joudaki, E., (2008). Optimization of caustic current efficiency in a zero-gap advanced chlor-alkali cell with application of genetic algorithm assisted by artificial neural networks. *Chemical Engineering Journal, 140*(1–3), 157–164.
72. Moussallem, I., Jörissen, J., Kunz, U., Pinnow, S., & Turek, T., (2008). Chlor-alkali electrolysis with oxygen depolarized cathodes: History, present status, and future prospects. *Journal of Applied Electrochemistry, 38*, 1177–1194.
73. Moussallem, I., Pinnow, S., Wagner, N., & Turek, T., (2012). Development of high-performance silver-based gas-diffusion electrodes for chlor-alkali electrolysis with oxygen depolarized cathodes. *Chemical Engineering and Processing: Process Intensification, 52*, 125–131.
74. Nickell, R. A., & Tatarchuk, B. J., (2005). High surface area supported precious metal cathodes utilizing metal microfibrous collectors for application in chlor-alkali cells. *Journal of Applied Electrochemistry, 35*(6).
75. O'Brien, T. F., Bommaraju, T. V., Hine, F., O'Brien, T. F., Bommaraju, T. V., & Hine, F., (2005). History of the chlor-alkali industry. *Handbook of Chlor-Alkali Technology: Volume I: Fundamentals, Volume II: Brine Treatment and Cell Operation, Volume III: Facility Design*.
76. O'Brien, T. F., Bommaraju, T. V., Hine, F., O'Brien, T. F., Bommaraju, T. V., & Hine, F., (2005). Chlor-alkali technologies. *Handbook of Chlor-Alkali Technology: Volume I: Fundamentals, Volume II: Brine Treatment and Cell Operation, Volume III: Facility Design and Product Handling, Volume IV: Plant Commissioning and Support Systems, Volume V: Corrosion, Environmental Issues, and Future Development*, 387–442.
77. Oloman, C., & Radcliffe, J., (1986). An experimental study of a fixed bed chlor-alkali reactor. *Journal of Applied Electrochemistry, 16*, 457–462.
78. Otashu, J. I., & Baldea, M., (2019). Demand response-oriented dynamic modeling and operational optimization of membrane-based chlor-alkali plants. *Computers and Chemical Engineering, 121*, 396–408.
79. Park, I. K., Ahn, C. Y., Lee, J. H., Lee, D. W., Lee, C. H., Cho, Y. H., & Sung, Y. E., (2019). Three-dimensionally interconnected titanium foam anode for an energy-efficient zero gap-type chlor-alkali electrolyzer. *International Journal of Hydrogen Energy, 44*(31), 16079–16086.
80. Perry, S. C., de León, C. P., & Walsh, F. C., (2020). The design,

performance, and continuing development of electrochemical reactors for clean electrosynthesis. *Journal of The Electrochemical Society, 167*(15), 155525.

81. Pletcher, D., & Pletcher, D., (1984). The chlor-alkali industry. *Industrial Electrochemistry*, 88–113.
82. Reig, M., Casas, S., Aladjem, C., Valderrama, C., Gibert, O., Valero, F., & Cortina, J. L., (2014). Concentration of NaCl from seawater reverse osmosis brines for the chlor-alkali industry by electrodialysis. *Desalination, 342*, 107–117.
83. Roh, K., Brée, L. C., Perrey, K., Bulan, A., & Mitsos, A., (2019). Optimal oversizing and operation of the switchable chlor-alkali electrolyzer for demand side management. In: *Computer Aided Chemical Engineering* (Vol. 46, pp. 1771–1776). Elsevier.
84. Roh, K., Brée, L. C., Perrey, K., Bulan, A., & Mitsos, A., (2019). Flexible operation of switchable chlor-alkali electrolysis for demand side management. *Applied Energy, 255*, 113880.
85. Roh, K., Brée, L. C., Perrey, K., Bulan, A., & Mitsos, A., (2022). On economic operation of switchable chlor-alkali electrolysis for demand-side management. In: *Simulation and Optimization in Process Engineering* (pp. 225–242). Elsevier.
86. Salah, A., Zhang, L., Tan, H., Yu, F., Lang, Z., Al-Ansi, N., & Li, Y., (2022). Advanced Ru/Ni/WC@ NPC multi-interfacial electrocatalyst for efficient sustainable hydrogen and chlor-alkali co-production. *Advanced Energy Materials, 12*(21), 2200332.
87. Sánchez-Sánchez, C. M., Expósito, E., Frias-Ferrer, A., González-García, J., Montiel, V., & Aldaz, A., (2004). Chlor–alkali industry: A laboratory scale approach. *Journal of Chemical Education, 81*(5), 698.
88. Schneiders, K., Zimmermann, A., & Henssen, G., (2001). Membrane electrolysis-innovation for the chlor-alkali industry. In: *Forum-Technische Mitteilungen Thyssenkrupp* (pp. 36–40). ThyssenKrupp.
89. Seko, M., (1976). The ion-exchange membrane, chlor-alkali process. *Industrial and Engineering Chemistry Product Research and Development, 15*(4), 286–292.
90. Shim, J. H., Jeong, J. Y., & Park, J. Y., (2015). Effects of operating conditions on improving alkali and electrical efficiency in chlor-alkali diaphragm cell. *Int. J. Electrochem. Sci., 10*, 6338–6347.
91. Špalek, O., (1994). Oxygen trickle-bed electrode as a cathode for chlor-alkali electrolysis. *Journal of Applied Electrochemistry, 24*(8), 751–757.

92. Stefan, P., Imad, M., & Thomas, T., (2010). Towards more energy-efficient chlor-alkali electrolysis. *Альтернативная Энергетика И Экология, 4*, 60–66.

93. Sugiyama, M., Saiki, K., Sakata, A., Aikawa, H., & Furuya, N., (2003). Accelerated degradation testing of gas diffusion electrodes for the chlor-alkali process. *Journal of Applied Electrochemistry, 33*, 929–932.

94. Tilak, B. V., Ramamurthy, A. C., & Conway, B. E., (1986). High performance electrode materials for the hydrogen evolution reaction from alkaline media. In: *Proceedings of the Indian Academy of Sciences-Chemical Sciences* (Vol. 97, pp. 359–393). Springer India.

95. Tonini, D. R., Gauvin, D. A., Soffel, R. W., & Freeman, W. P., (2003). Achieving low mercury concentrations in chlor-alkali wastewaters. *Environmental Progress, 22*(3), 167–173.

96. Trasatti, S., (1992). Electrocatalysis of hydrogen evolution: Progress in cathode activation. *Advances in Electrochemical Science and Engineering, 2*, 1–85.

97. Van Zee, J. W., (1998). *Advances in Mathematical Modeling and Simulation of Electrochemical Processes and Oxygen Depolarized Cathodes and Activated Cathodes for Chlor-alkali and Chlorate Processes*. The Electrochemical Society.

98. Vorce, L. D., (1944). Historic development of caustic-chlorine cells in America. *Transactions of the Electrochemical Society, 86*(1), 69.

99. Yamaguchi, K., Senda, A., & Sakata, A., (1990). A nickel-based composite electroplated cathode for the membrane-type chlor-alkali cell. *Journal of the Electrochemical Society, 137*(5), 1419.

100. Yoshida, N., Yoshitake, M., Endoh, E., & Morimoto, T., (1989). Development of highly durable low hydrogen overvoltage cathode in chlor-alkali cells. *International Journal of Hydrogen Energy, 14*(2), 137–140.

101. Zhang, L. N., Lang, Z. L., Wang, Y. H., Tan, H. Q., Zang, H. Y., Kang, Z. H., & Li, Y. G., (2019). Cable-like Ru/WNO@ C nanowires for simultaneous high-efficiency hydrogen evolution and low-energy consumption chlor-alkali electrolysis. *Energy and Environmental Science, 12*(8), 2569–2580.

102. Zhang, L. N., Li, R., Zang, H. Y., Tan, H. Q., Kang, Z. H., Wang, Y. H., & Li, Y. G., (2021). Advanced hydrogen evolution electrocatalysts promising sustainable hydrogen and chlor-alkali co-production. *Energy and Environmental Science, 14*(12), 6191–6210.

103. Zhang, Y., Liu, X., Li, W., Liu, W., Yin, H., & Wang, D., (2023). Advanced and durable self-standing MoC–Mo_2C electrode for alkaline hydrogen

evolution in chlor-alkali electrolysis. *ACS Sustainable Chemistry and Engineering*.

104. Zhao, A., Zhong, F., Feng, X., Chen, W., Ai, X., Yang, H., & Cao, Y., (2019). A membrane-free and energy-efficient three-step chlor-alkali electrolysis with higher-purity NaOH production. *ACS Applied Materials and Interfaces, 11*(48), 45126–45132.
105. Zheng, N., Liu, J., Wang, Q., & Liang, Z., (2011). Mercury contamination due to zinc smelting and chlor-alkali production in NE China. *Applied Geochemistry, 26*(2), 188–193

CHAPTER

6 Chemical Engineering Principles in Chlor-Alkali Industry

Contents

6.1. INTRODUCTION

The aim of this chapter is to synthesize key elements related to the engineering methodologies and unit operations that are relevant to the chlor-alkali process. The introductory section of the chapter involves an analysis of energy and material balances. The concepts mentioned above hold a fundamental position in the discipline of chemical engineering and are implicitly utilized throughout the entire body of literature in this field. Within this particular context, we shall present several essential concepts. This bears considerable significance within the domain of electrochemical processing. The presentation scrutinizes a range of methodologies for predicting and determining the present dispersion in heterogeneous electrochemical systems (Hine et al., 1986).

The current approach is succinct and endeavors to provide a comprehensive perspective. Numerous textbooks and scholarly literature review publications offer a wealth of comprehensive information. The current chapter refrains from delving into the practical intricacies pertaining to engineering and operation.

Did You Know?

Chlorine is a highly reactive element and was first isolated by the Swedish chemist Carl Wilhelm Scheele in 1774.

6.2. MATERIAL AND ENERGY BALANCE

Within the field of chemistry, it is customary to incorporate a designated input material into a reaction vessel for the purpose of undergoing transformation, leading to the generation of a targeted output with restricted kinetics and efficacy. The preparatory procedure is frequently employed to retrieve any residual feedstock from the exhausted stream. Apart from the primary resources, the manufacturing procedure also encompasses the utilization of energy and utilities (White, 1992). The inclusion of both the sensible and latent heat of the raw materials and final products is a crucial aspect in determining the overall energy balance. The product mentioned above is subjected to segregation from the waste stream and subsequently subjected to additional processing in the finishing section to enhance its transportability, as depicted in Figure 6.1(A) (Otashu & Baldea, 2019). The energy balance in chemical processing is intricately connected to the transfer of substances within the examined flowchart. Figure 6.1(B) provides a comprehensive representation of the energy transfer mechanisms occurring both inside and outside the reactor. The analysis of the combined material and energy movement in the three color-alkali methods demonstrates a significant exchange of energy among the different unit operations (Slater & Hesketh, 2004).

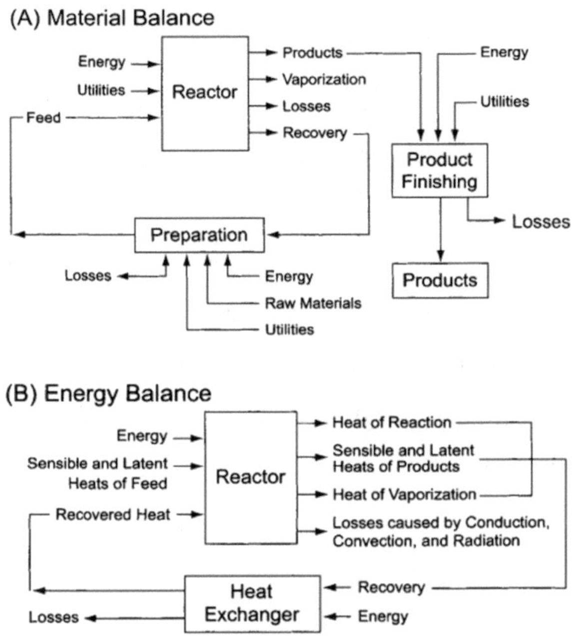

Figure 6.1. *Illustration of material and energy balance around a reactor.*

Source: Fumio Hine, Creative Commons License.

The upcoming section, namely Section 6.2.1, will provide a concise overview of the fundamental concepts pertaining to thermodynamics. The principle of energy balance is rooted in these concepts. The previously mentioned ideas will be utilized in a subsequent fashion for an electrochemical procedure, as elaborated in Section 6.2.2.

6.2.1. Thermodynamics

6.2.1.1. Energy Changes Initiated by Chemical Reactions

The amount of energy supplied to a particular system can be denoted as Q, which comprises the total of the system's work performed (If) and changes in its internal energy (U). The aforementioned assertion pertains to the fundamental principle commonly acknowledged as the primary law of thermodynamics. The distinctive feature of electrochemical energy, denoted as nFE, is depicted in Figure 6.2, which sets it apart from other types of energy input (Liberman et al., 2012).

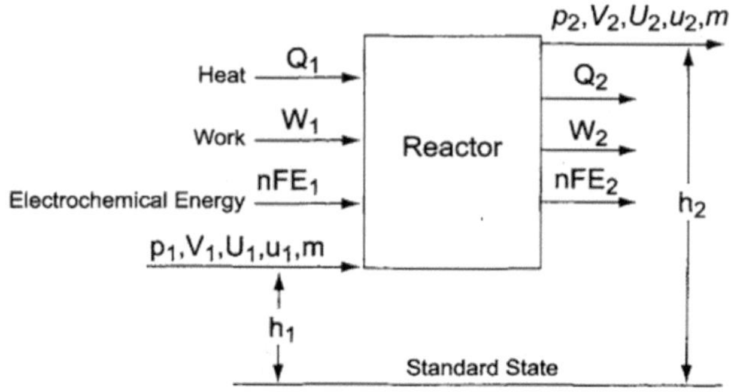

Figure 6.2. *Illustration of an energy balance around a continuous reactor.*

Source: Robert A. Arnold, Creative Commons License.

The following equations delineate the energy outflow and inflow that traverse a continuous reactor.

$$\text{Energy inflow} = Q_1 - W_1 - nFE_1 + h_1\frac{g}{g_c} + p_1V_1 + \frac{u_1^2}{2g_c} + U_1$$

$$\text{Outflow energy} = -Q_2 + W_2 + nFE_2 + h_2\frac{g}{g_c} + p_2V_2 + \frac{u_2^2}{2g_c} + U_2$$

nFE = electrochemical energy,

$h\frac{g}{g_c}$ = potential energy,

pV = work done by pressure,

and $\frac{u^2}{2g_c}$ = kinetic energy.

Since the energy input and output are equal, we have the following:

$$Q - W - nFE = \Delta H + \frac{(h_2 - h_1)g}{g_c} + \frac{u_2^2 - u_1^2}{2g_c}$$

where;

$Q = Q_1 + Q_2$

$W = W_1 + W_2$

$E = E_1 + E_2$

$H_i = p_iV_i + U_i$ (enthalpy)

$\Delta H = H_2 - H_1$ (enthalpy change)

6.2.1.2. Sensible and Latent Heat

The equivalence between the enthalpy change (ΔH) and the heat influx resulting from a constant pressure-volume change is expressed by the following equation, which represents the specific heat:

$$C_p = \left(\frac{\partial H}{\partial T}\right)_p$$

or

$$\Delta H = \int_{T_1}^{T_2} C_p dT$$

It is a variable that is dependent on temperature and can be expressed as:

$$C_p = a_0 + a_1 T + a_2 T^2 + + a_3 T^3 ... + + a_{-2} T^{-2}$$

The standard thermodynamic information compilations contain the coefficients (Zhang et al., 2014). The enthalpy change associated with the transition from one state to another of water from solid to gas, while passing through the liquid phase, is represented as follows:

$$\Delta H = \int_{T_1}^{T_m} C_p(s)\, dT + \lambda_m + \int_{T_m}^{T_b} C_p(l)\, dT + \lambda_b + \int_{T_b}^{T_2} C_p(g)\, dT$$

The equation utilizes several symbols, specifically "Where" to indicate the latent heat of fusion, "latent heat of vaporization" to represent the latent heat of vaporization, "s" to signify the solid phase, "(I)" to indicate the liquid phase, "(g)" to represent the gaseous phase, and "1" and "2" as subscripts to denote the initial and final states correspondingly.

Furthermore, it should be noted that the variable "m" is utilized to denote the melting point, while the variable "b" is employed to indicate the boiling point.

The aforementioned equation contains integration terms that correspond to the sensible heats. The Clausius Clapeyron equation is utilized to determine latent heat (Hine & Hine, 1985).

$$l = T(V_2 - V_1)\frac{dp_i}{dT}$$

The volumetric measurements before and after the phase transition are denoted as V_1 and V_2, respectively, whereas the pressure exerted by the vapor is indicated by the symbol.

6.2.1.3. Isothermal Change and Adiabatic Change

Under isothermal conditions, the internal energy of a system remains constant, meaning that the change in internal energy (ΔU) is equal to zero. Thus, when considering one mole of an ideal gas, the internal energy remains unchanged:

$$Q = W = RT \ln\left(\frac{V_2}{V_1}\right) = RT \ln\left(\frac{p_1}{p_2}\right)$$

In the context of an adiabatic process, there is a lack of heat transfer, resulting in Q being equal to zero. Hence,

$$\Delta U = -W = -pV$$

After substituting,

$$C_p = C_v + R$$

$$\gamma = C_p / C_V$$

$$\frac{T_2}{T_1} = \left(\frac{V_1}{V_2}\right)^{\gamma-1} = \left(\frac{p_2}{p_1}\right)^{(\gamma-1)/\gamma}$$

$$pV^{\gamma} = \text{constant}$$

The mathematical representation representing the complete quantity of work, denoted as W, executed throughout the cycle can be expressed as follows (Figure 6.3):

$$W = W_{AB} + W_{BC} + W_{CD} + W_{DA}$$

The subscripts denote the corresponding stages.

$$W_{BC} = -U_{BC} = -\int_{T_1}^{T_2} C_V \, dT$$

$$W_{DA} = -U_{DA} = -\int_{T_1}^{T_2} C_V \, dT = -W_{BC}$$

$$\frac{p_C}{p_D} = \frac{p_B}{p_A}$$

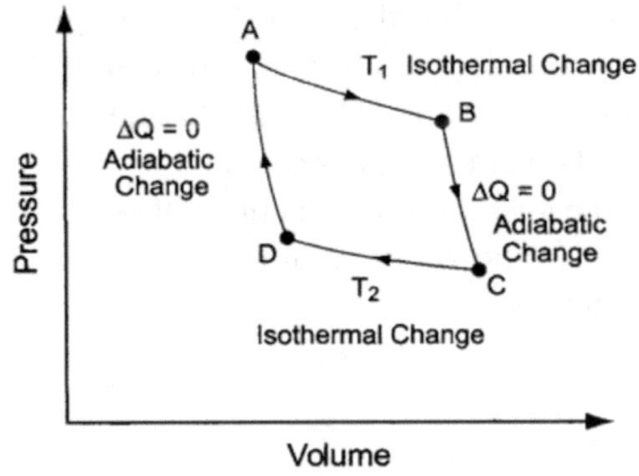

Figure 6.3. *The depiction of Camot's cycle for an ideal gas.*

Source: Thomas F. O'Brien, Arnold, Creative Commons License.

Upon substitution of the above equations, the resulting outcome is:

$$W = RT_1 \ln\left(\frac{p_B}{p_A}\right) + RT_2 \ln\left(\frac{p_C}{p_D}\right)$$

$$= RT_1 \ln\left(\frac{p_B}{p_A}\right)\left(1 + \frac{T_2 \ln(p_C/p_D)}{T_1 \ln(p_B/p_A)}\right)$$

$$\frac{W}{Q_1} = 1 - \frac{T_2}{T_1}$$

$$Q_1 = W_{AB} = RT_1 \ln(p_B/p_A)$$

The equation signifies that the variable Q denotes the heat inflow, whereas the variable W represents the work accomplished by the surroundings. The thermodynamic parameter W/representing the energy efficiency of a heat engine is contingent upon temperature but remains unaffected by the condition of the surrounding environment. Consequently, by using the equations mentioned above, it can be demonstrated that (Karlova et al., 2016).

$$\frac{Q_1}{T_1} + \frac{Q_2}{T_2} = 0$$

or

$$\oint \frac{dQ\,(R)}{T} = 0$$

Thus, it can be inferred that the overall alteration in the state variable over the course of a complete cycle is equal to null. In other words, the state functions pertaining to steps AB and CD demonstrate changes in ΔU and ΔV, respectively. The equations presented above illustrate that the state function of a reversible system remains constant regardless of the pathway. The entropy variation is the recently introduced thermodynamic property (Tishin & Spichkin, 2002).

$$dS = \frac{dQ\,(R)}{T}$$

or

$$\Delta S = \int \frac{dQ\,(R)}{T}$$

Figure 6.4 depicts the phenomenon of heat transfer occurring between two adjacent rooms. The process of heat transfer, denoted as Q, occurring between a system operating at a temperature T and its surrounding environment operating at a different temperature leads to a consequential alteration in the entropy of the system, represented as S.

$$\Delta S = \left(\frac{Q}{T_2} - \frac{Q}{T_1}\right) = \frac{Q}{T_2}\left(1 - \frac{T_2}{T_1}\right)$$

$$W = T_2\,\Delta S$$

Work loss can be expressed as:

$$S - S_s = \int_{T_s}^{T_m} \frac{C_p(s)}{T}\,dT + \frac{\lambda_m}{T} + \int_{T_m}^{T_b} \frac{C_p(l)}{T}\,dT + \frac{\lambda_b}{T} + \int_{T_b}^{T} \frac{C_p(g)}{T}\,dT$$

The value of S can be computed utilizing the aforementioned equation, given that the value is already established (Crossley et al., 2021).

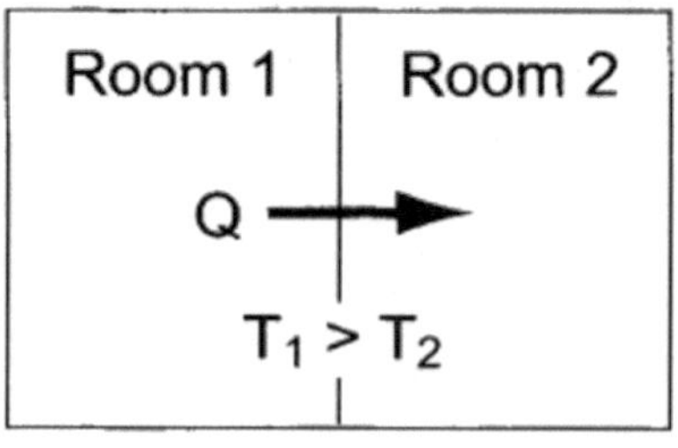

Figure 6.4. *Illustration of heat transfer between rooms and entropy change.*

Source: T. Sugino, Arnold, Creative Commons License.

6.2.1.4. Free Energy and Chemical Equilibrium

It is possible to summarize additional state functions, specifically the Helmholtz free energy (A) and the Gibbs free energy (G), which have been previously discussed.

$$A = U - TS$$

$$G = H - TS = U + pV - TS = A + pV$$

$$\mathrm{d}U = \mathrm{d}Q - dW = T\,\mathrm{d}S - p\,\mathrm{d}V$$

$$\mathrm{d}A = \mathrm{d}U - T\,\mathrm{d}S - S\,\mathrm{d}T = -[p\,\mathrm{d}V - S\,\mathrm{d}T]$$

or

$$\Delta A = \Delta U - T\Delta S$$

and

$$\mathrm{d}G = \mathrm{d}H - T\,\mathrm{d}S - S\,\mathrm{d}T = V\,\mathrm{d}p - S\,\mathrm{d}T$$

As per the principles of thermodynamics, the reversible process heat generated in a is denoted by Tk S, while the work by A = ( ). In practical applications, it is observed that the system does not exhibit ideal reversibility, resulting in the work done on the surroundings being consistently less than the value of A. Stated differently, A represents the upper limit of work attainable during an isothermal process (Hu & Yang, 2008).

Given the aforementioned context, one may contemplate an electrolytic cell comprising Zinc and platinum sheets immersed in a solution of diluted sulfuric acid, wherein the two electrodes are externally connected. The ensuing chemical transformation takes place:

$$\underset{\text{(state 1)}}{Zn + 2H^+} = \underset{\text{(state 2)}}{Zn^{2+} + H_2}$$

The process yields a production of 2 Faradays of direct current electricity per mole of Zinc dissolved. The inverse procedure can be initiated by the provision of external energy input. Whilst DC electricity is produced through a galvanic cell (as per Eqn. (35)), the energy resulting from the alteration in volume (pV) cannot be utilized for alternative purposes (Ensing et al., 2006). Hence, the utmost amount of work that can produce energy is.

$$W_E = -\Delta A - p\Delta V = -\Delta G$$

The aforementioned equation is equivalent to nFE. Therefore, the highest potential difference or the electromotive force (EMF) exhibited by the galvanic cell is:

$$E_F = -\frac{\Delta G}{nF}$$

Due to the exothermic nature of the forward reaction, entropy change gives a positive value. Consequently, the Gibbs free energy (G) must be negative. The impact of alterations in temperature at constant pressure on the Gibbs free energy change can be expressed as follows:

$$\left(\frac{d\Delta G}{dT}\right)_p = -\Delta S = -\frac{\Delta H - \Delta G}{T}$$

The Gibbs-Helmholtz equation, along with other pertinent electrochemical relationships, can be derived from the aforementioned pair of equations.

$$E + \frac{\Delta H}{nF} = T\frac{dE}{dT}$$

$$\frac{d\left(\frac{\Delta G}{T}\right)}{dT} = -\frac{\Delta H}{T^2}$$

$$\frac{d\left(\frac{\Delta G}{T}\right)}{d\left(\frac{1}{T}\right)} = \Delta H$$

In batch and continuous modes:

$$\Delta U_R = \Sigma U_F - \Sigma U_I$$

$$\Delta H_R = \Sigma H_F - \Sigma H_I$$

The symbols I and F are employed to denote the initial and final states, respectively. The equations mentioned earlier serve as an example of the Hess law of constant summation, which offers an alternative representation of the first law of thermodynamics. The determination of reaction heat can be accomplished through the utilization of these equations, given that the enthalpies of both states are provided. The symbols < 0 and > 0 are utilized in denoting exothermic and endothermic reactions, correspondingly (Muller & Warshel, 1995). The Gibbs free energy or chemical potential of a substance that has achieved equilibrium under the conditions of temperature T and pressure p can be represented by the following equation.

$$\mu = G = H - TS$$

Applying thermodynamics laws:

$$\Delta H = Q = T\Delta S$$

It follows that:

$$\Delta\mu = \Delta H - T\Delta S = 0$$

Hence,

$$\Sigma\mu = 0$$

The chemical potential is a term commonly used to denote the Gibbs free energy of one mole of a specific species. The term in question is commonly utilized within the context of chemical processes owing to its advantageous nature. The chemical capacity of a gaseous species is defined by considering its reliance on temperature and pressure.

$$\mu = \mu^o + RT\ln f$$

The variable f denotes the fugacity, which is an effective pressure, while $\mu°$ represents the standard chemical potential at f = 1 (Kollman et al., 2001).

It is significant that these expressions hold for liquids or solutions as well, provided that the activity, a, is utilized as a substitute for f:

$$\mu = \mu^o + RT\ln a$$

where;

$$a = \gamma C$$

In the given context, γ represents the activity coefficient, and C represents the concentration. It is conventionally observed that γ approaches unity as the solution becomes infinitely diluted (i.e., $C \rightarrow 0$). When the aforementioned equation is applied to the chemical equilibrium process:

$$\mu_A^o + \mu_B^o + \cdots + RT\ln(a_A \cdot a_B \cdots) = \mu_C^o + \mu_D^o + \cdots + RT\ln(a_C \cdot a_D \cdots)$$

Expressing in an alternate form:

$$\Delta G^o = -RT\ln\frac{\Pi(a_F)}{\Pi(a_I)}$$

where;

$$\Delta G^o = \sum \mu_F^o - \sum \mu_I^o = (\mu_C^o + \mu_D^o + \cdots) - (\mu_A^o + \mu_B^o + \cdots)$$

$$\Pi(a_I) = a_A \cdot a_B \cdots$$

$$\Pi(a_F) = a_C \cdot a_D \cdots$$

Since,

$$K = \frac{a_C \cdot a_D \cdots}{a_A \cdot a_B \cdots} = \frac{\Pi(a_F)}{\Pi(a_I)}$$

We have:

$$\Delta G^o = -RT \ln K$$

6.2.2. Energy Balance in Electrochemical Processes

6.2.2.1. Thermodynamic Decomposition Voltage and Thermoneutral Voltage

It is indisputable that electricity serves as the predominant energy source provided to an electrochemical reactor. The chemical reaction within the electrolyzer is facilitated by the application of thermal energy, which promotes its progression. Based on empirical data, it has been observed that the electric current required by the cell exceeds the value predicted by Faraday's law. This is due to the occurrence of secondary reactions that utilize a portion of the electricity, leading to a reduction in current efficiency (Oloman & Li, 2008).

The energy demands of a cell utilizing electrolysis exhibit a direct correlation with the voltage of the cell while displaying an inverse correlation with the efficiency of the cell. The present discourse pertains to the determinants that regulate the contemporary efficacy and electrical potential of the cellular unit.

The fundamental basis of this concept pertains to alterations in the free energy of a given reaction, which can be mathematically represented as follows:

$$E_{th}^o = -\frac{\Delta G^o}{nF}$$

Thus, for the reaction:

$$2NaCl + 2H_2O \rightarrow 2NaOH + H_2 + Cl_2$$

The values of ΔG°, determined based on the available free energy data, are documented in Table 6.1, alongside the corresponding E^{oth} values at different temperatures. It is worth noting that the reaction mentioned above is characterized by an endothermic nature, and Table 6.1 provides details about the overall heat involved in this particular reaction process (Bernardi et al., 1985).

Table 6.1. Free Energy and Enthalpy Change at Various Temperatures

Reaction	**Tem-pera-ture (°C)**	**G° (Kcal)**	**H°**	**(V)**	
2NaCl + O 2NaOH + +	80	100.05	105.09	2.169	2.278
	90	99.91	104.81	2.166	2.272
	100	99.77	104.53	2.163	2.266
2KCl + O—+ 2KOH + +	80	100.05	105.09	2.169	2.278
	90	99.91	104.81	2.166	2.272
	100	99.77	104.67	2.163	2.269
0.5 2 + 2NaCl +O 2NaOH +	80	45.47	37.19	0.986	0.806
	90	45.71	36.99	0.991	0.802
	100	45.96	36.78	0.996	0.797
HCl 0.5 + 0.5	25	40.02	31.35	1.36	1.74

Source: K. Aoki, Creative Commons License.

$$E^{o}_{tn} = -\frac{\Delta H}{nF}$$

The thermo-neutral voltage, denoted as E, is defined as the voltage that corresponds to the heat of the reaction, at which neither heat is lost to the surroundings nor wanted by the electrolytic cell.

The importance of thermoneutral voltage can be illustrated in this particular scenario by conducting a comparative examination of the electrolysis of HCl solutions, taking into account their corresponding voltage values. As shown in Table 6.1, is always greater than because the term contains the heat associated with the entropy change for the reaction 2HCl + If a cell operates at a voltage between E and ñ , the cell will be cooled as it dissipates the heat corresponding to the entropy change, irreversibly

(Panigrahi & Dubey, 2019). On the other hand, if exposed to voltages exceeding a specific threshold, the cell encounters a rise in temperature as a result of the excess energy required to overcome the overvoltages and ohmic losses. As a result, it is imperative to implement cooling measures in order to uphold a pre-established temperature. The formula provided can be utilized to calculate the thermal energy generated or utilized by the system (Macdonlad & Challingsworth, 1993). The amount of thermal energy released is denoted as:

$$Q_{rev} = -T\Delta S = \Delta G - \Delta H$$
$$= nFE^{o}_{th} - \Delta H$$

Cell voltage is expressed as:

$$E = E^{o}_{th} + \Sigma IR + \Sigma \eta$$

where; h represent the overvoltage, I R represents the ohmic drop.

The irreversible heat generated $Q_{irr} - Q_{rev}$ is given by:

$$Q_{irr} = (E - E^{o}_{th})nF$$

$$Q = nFE - \Delta H$$

In cases where Q is positive, the system releases heat, whereas, in instances where Q is negative, the system absorbs heat.

The voltage of the following reaction in chlor-alkali cells is 2.166 V at a temperature of 90°C and 2.272 V at another temperature. Empirically, the cells function within the voltage range of 3.3–3.5 V (Rose et al., 1948). Consequently, the quantity of thermal energy produced is:

$$Q = \left[\left(\frac{100}{CE}\right)(46.099E)\right] - \Delta H$$

For 96% current efficiency and voltage of 3.3 V:

$$Q = \frac{100}{96}(46.099(3.3)) - 104.81 = 158.46 - 104.81$$
$$= 53.655\,\text{kcal mol}^{-1}\text{Cl}_2 \text{ or } 0.7566\,\text{kcal g}^{-1}\text{Cl}_2$$
$$= 3165\,\text{kJ kg}^{-1}\text{Cl}_2$$

6.2.2.2. Voltage and Energy Balances

Chlor-alkali cells' energy demands are presented in Figures 6.5–6.7.

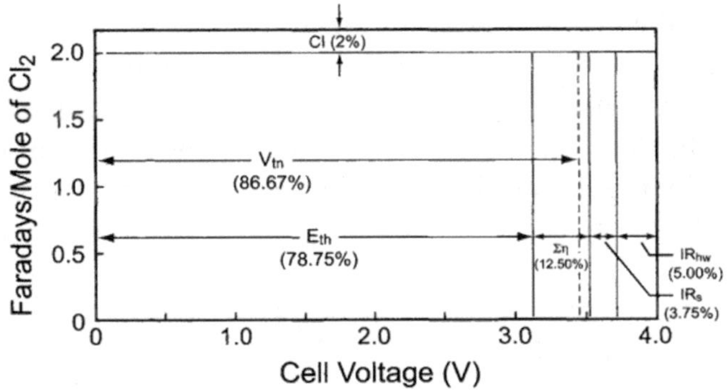

Figure 6.5. *Representation of energy distribution in an amalgam chlor-alkali cell (DeNora 24M-2) operating at 10 kA m^{-2}.*

Source: T. R. Beck, Creative Commons License.

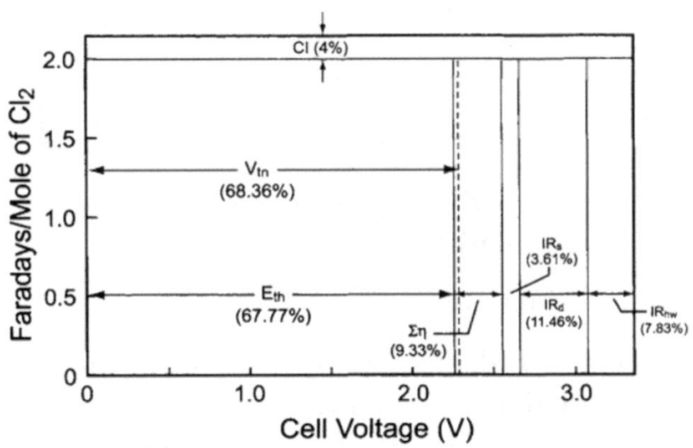

Figure 6.6. *Representation of energy distribution in a diaphragm chlor-alkali cell (ELTECH H-4/75) operating at 2.3 kAm^{-2}.*

Source: C. T. Brown, Creative Commons License.

The figure displays the existing inefficiencies, denoted as CI, at the highest point, while the voltage distribution is depicted in its constituent parts alongside the thermoneutral voltage that pertains to the specific circumstances. Table 6.1 presents the H and G values pertaining to the reaction. The pertinent information pertaining to mercury cells is determined through the utilization of the Gibbs Helmholtz equation. The E° and values are computed based on the data presented in the reference, while the G values are obtained from the reference source. The energy profile of an amalgam-type chlorine cell is illustrated in Figure 6.5. It can be observed that a portion of the heat produced due to the irreversible losses is utilized to maintain the temperature of the cell at a predetermined level. It is evident from Figure 6.5 that there exists additional heat that requires removal. A portion of the thermal energy is extracted from the amalgam that is discharged from the electrolytic cell. Consequently, it is feasible to achieve thermo-neutrality in mercury cells by regulating the temperature and current density of the cell (Yeetsorn et al., 2020).

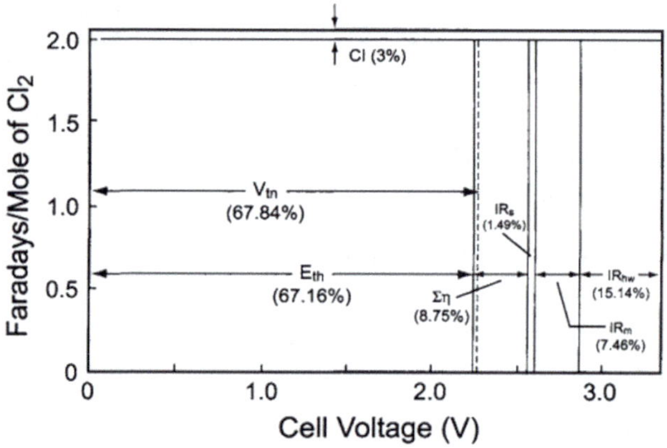

Figure 6.7. *Representation of energy distribution in a membrane chlor-alkali cell (MGC-26) operating at 5 kA m^{-2}.*

Source: David Francis, Creative Commons License.

Figures 6.6 and 6.7 depict energy diagrams that demonstrate the proximity of the and the in both the membrane and diaphragm cells. Consequently, despite notable progress in membrane technology and hardware advancements, their operation at the aforementioned level remains exceedingly arduous.

It is noteworthy that the cathode-based membrane cells utilizing air depolarization exhibit a voltage of 0.991 V at 90°C in comparison to a voltage of 0.802 V (Kugler et al., 2014).

Remember:

Sodium hydroxide is a versatile chemical used in various industries, including paper manufacturing, textiles soaps and detergents.

The current definition of thermo-neutral voltage fails to account for the energy required to raise the temperature of the feed stream from the standard temperature of 25°C to the operating temperature of the electrolyte and to fully saturate the product gases with water vapor. To address this, a revised thermo-neutral voltage, known as the thermo-balance voltage, has been proposed. Considering the aforementioned heat requirements, the thermo-balance voltage has been established as a practical solution for cells employed in water electrolysis (Chavan et al., 2018). The voltage of the thermo-balance is formally characterized as:

$$V_{tb} = V_{mtn} + V_{rad} + V_{conv}$$

$$V_{rad} = A\varepsilon s\,(T^4 - T^4_{\ a})10^{-3} / I$$

$$V_{conv} = 1.77A(T - T_a)^{1.25}10^{-3} / I$$

where A is the radiating area (in), s is the emissivity, cr is the Stefan Boltzmann constant (5.67 x $W m^2 K^4$)• a is the ambient temperature in K, is the operating cell temperature in K, and I is the load in kA.

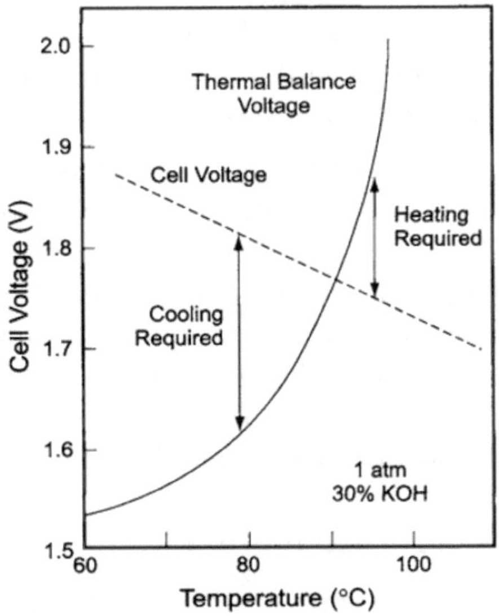

Figure 6.8. *Illustration of isothermal and thermal-balance voltages of a water electrolyzer as a function of temperature at 1 atm total pressure.*

Source: Adam Jacobsen, Creative Commons License.

The outcomes of the aforementioned computations for water electrolysis cells are illustrated in Figure 6.8. The graph depicts the variation of cell voltage with respect to temperature, wherein the discontinuous line intersects the thermal-balance voltage curve at approximately 90°C. At temperatures below 90°C, the voltage of the cell exceeds both the thermo balance voltage and the voltage that corresponds to the overall heat requirement.

Consequently, the cell releases surplus heat, which necessitates its removal. Conversely, when the temperature exceeds 90°C, the amount of heat produced is inadequate to offset the overall heat demand of the reaction, necessitating the heating of the cell. Consequently, the cell has the ability to function at a temperature of approximately 90°C without the need for any external heating or cooling mechanisms (Hine, 1985).

6.3. CURRENT DISTRIBUTION

Non-uniform distributions in chlor-alkali cells may lead to unfavorable electrode reactions, operational corrosion of cell components, and localized blister formation due to membrane drying. Non-uniform current distribution during electroplating processes can result in various issues, such as uneven plating, the formation of dendritic metal growth, and the deposition of impurities on cathodes. The cumulative impact of these factors ultimately culminates in substandard product quality and heightened energy consumption (Newman & Tobias, 1962).

The groundbreaking work by Kasper, Wagner, and other researchers laid the mathematical groundwork for calculating the distribution of potential and current. In this discussion, a concise summary of these fundamental principles and their application in chlor-alkali operations is presented. The distribution of electrochemical systems is currently influenced by factors such as cell geometry, electrode kinetics, and mass transfer effects. The resulting current distributions, arising from specific factors, are categorized as primary, secondary, and tertiary contributions (Fludapanjeh et al., 2019).

6.3.1. Primary Current Distribution

The determination of electric current distribution is predominantly reliant on the geometry of the cell, provided that the electrode maintains a uniform potential and any effects of activation overpotential and mass transfer can be neglected. The standard methodology for ascertaining this entails resolving the Laplace equation within the volume while imposing either Neumann or Dirichlet boundary constraints. The Dirichlet boundary condition pertains to the determination of the electric potential at the electrode, while the Neumann boundary condition is concerned with the computation of the flux on the surface. The Laplace equation solution is responsible for determining the potential distribution, whereas the primary current distribution is obtained by differentiating the potential distribution along the selected direction, which is usually perpendicular to the electrode surface (Prentice & Tobias,

1982). In this particular context, we shall examine a scenario that entails a set of parallel plates that are upheld at potentials denoted as V_1 and V_2, correspondingly, as depicted in Figure 6.9(A). The Laplace equation can be expressed mathematically in the following manner.

Given the absence of end effects, the boundary conditions can be expressed as follows:

$$V = \frac{d^2V}{dx^2} + \frac{d^2V}{dy^2} = 0$$

$$V = V_1 \quad \text{at } x = 0$$

$$V = V_2 \quad \text{at } x = d$$

The solution of the equations mentioned above results in…

$$V = \frac{V_0 - V_1}{d}x + V_1$$

Since,

$$i = -k \text{ grad } V = k\frac{dV}{dx}$$

$$i = \frac{k}{d}(V_0 - V_1)$$

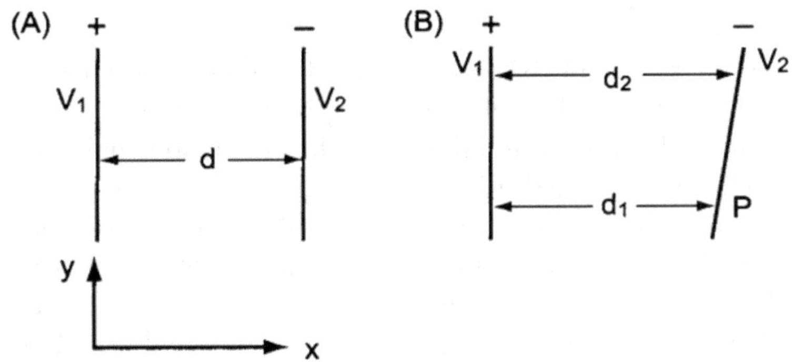

Figure 6.9. *Schematic of a cell with parallel plates (A); and with an inclined cathode (B).*

Source: Steve Brien, Creative Commons License.

The symbol "k" represents the conductivity of the medium, while "i" denotes the current density. Upon modifying the electrode arrangement to resemble that depicted in Figure 6.9(B), such as in the Hull cell utilized for electroplating experiments, it can be demonstrated that the current density at a specific location on the slanted cathode is directly proportional to the reciprocal of the distance separating the electrodes. It is noteworthy that electric current has a tendency to accumulate at locations with acute angles. In the event that the electrode is configured in a saw-tooth pattern, the density of current is anticipated to be elevated at the highest point and comparatively lower within the trough (Newman & Tiedermann, 1993).

The aforementioned instances pertain to uncomplicated geometrical shapes. As the complexity of the electrode geometry increases, the potential distribution becomes more intricate. Furthermore, the resolution of the Laplace equation and the attainment of an analytical solution pose significant challenges. This necessitates a numerical solution to the problem.

The following discourse pertains to the analysis of the two-dimensional field, as depicted in Figure 6.10. The Laplace equation and its corresponding boundary conditions can be expressed as follows:

$$\frac{\partial^2 V}{\partial x^2} + \frac{\partial^2 V}{\partial y^2} = 0$$

$$V = 0 \quad \text{at } x = 0$$

$$V = V_0 \quad \text{at } x = a$$

$$\frac{\partial V}{\partial y} = 0 \quad \text{at } y = 0$$

The variables "a" and "b" correspond to the inter-electrode distance and the cell width, respectively. Solving the Laplace equation using the Fourier series becomes relatively straightforward when the field is unbounded in all directions. However, when dealing with the standard setup of parallel-planar electrodes positioned in a rectangular chamber with insulating barriers surrounding them, the mathematical analysis becomes more intricate (Georgiadou, 2003). The computation of a two-dimensional electrode field with a relatively uncomplicated geometry can be achieved through the utilization of a conformal transformation that is rooted in the principles of complex functions. In this context, we shall examine a pair of intricate variables, denoted as z and w, where the latter is reliant on the former as a function.

$$z = x + jy$$

$$w = u + jv$$

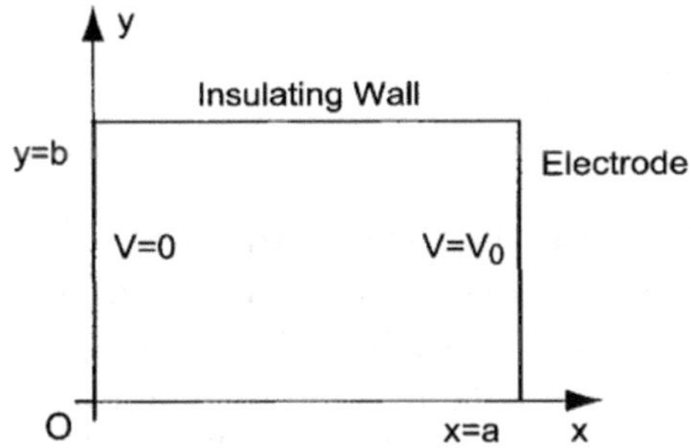

Figure 6.10. *Illustration of a rectangular cell with parallel flat-plate electrodes.*

Source: Roger E. Shamel, Creative Commons License.

Where "j" represents the imaginary unit. Upon substitution of the following equations into the Laplace equation, the resulting expression is:

$$\frac{\partial u}{\partial x} = \frac{\partial v}{\partial y}, \ \frac{\partial u}{\partial y} = -\frac{\partial v}{\partial x}$$

$$\frac{\partial^2 u}{\partial x^2} + \frac{\partial^2 u}{\partial y^2} = \frac{\partial^2 v}{\partial x^2} + \frac{\partial^2 v}{\partial y^2} = 0$$

Figure 6.11 presents a detailed graphical representation of both the z-plane and w-plane coordinates. It is important to note that the curve depicted on the w-plane is directly related to the curve on the z-plane, as it depends on the function of z. Consequently, any line or figure on the z-plane undergoes a conformal transformation when mapped onto the w-plane. Interestingly, sets of lines that maintain a constant value of ii and U can intersect each other at right angles. Moreover, the transformed curves on the z-plane also exhibit orthogonality. To elaborate further, the sets of contours on the complex plane z correspond to specific u and v values, representing the flow lines and equipotential lines (He & Li, 2000).

Consider the following equation as an illustrative example of a conformal transformation.

$$w = \log z$$

or

$$u + jv = 0.5 \log(x^2 + y^2) + j \tan^{-1}(y/x)$$

Therefore,

$$(x^2 + y^2) = \exp(2u)$$

$$y/x = \tan v$$

A linear locus in the in-plane, denoted by the variable u, can be associated with a circular path in the z-plane. Similarly, a radial locus in the in-plane, denoted by the variable v, can be linked to a straight line in the z-plane. The diagram illustrates that a decrease in diameter results in a reduction of the gap between adjacent circles, which implies an increase in the potential gradient.

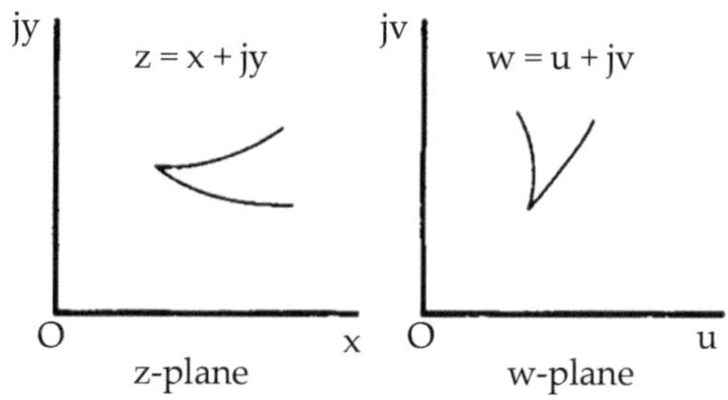

Figure 6.11. *Representation of conformal transformation between the z-plane and the w-plane.*

Source: Janet L. White, Creative Commons License.

The conformal transformation is a mathematical method used to calculate complex potential fields by mapping two holomorphic functions, z and in, as mentioned earlier. The potential field in the z-plane can be determined by acquiring a function represented as *f*(z). In situations where the specific function describing the potential field is unknown, the Schwarz-Christoffel transformation can be applied for analysis, as explained here. This mapping function converts the interior of a convex polygon in the z-plane to an intermediate function or plane positioned in the upper region of the complex plane (Datta & Harris, 1997).

$$z = C_1 \int \prod_{i=1}^{n} (\xi - \xi_1)^{-\lambda_i} d\xi + C_2$$

$$\sum_{i=1}^{n} \lambda_i = 2$$

C1 and C2 are constants, and theta represents the external angle at vertex O of the polygon.

The correlation between the variables e and w is:

$$z - ja = \frac{a}{\pi}(\xi - \log \xi)$$

$$w - jV = -\frac{V}{\pi}\log \xi$$

6.3.2. Secondary Current Distribution

If the potential is uniform across the electrode surface, the main current distribution is solely impacted by the cell geometry. The phenomenon of secondary current distribution is influenced by the kinetics of the electrode and is incorporated as a boundary condition at the surface of the electrode in the Laplace equation. At the present level of the distribution description, mass transfer effects are not taken into consideration (Liu et al., 2004) (Figure 6.12).

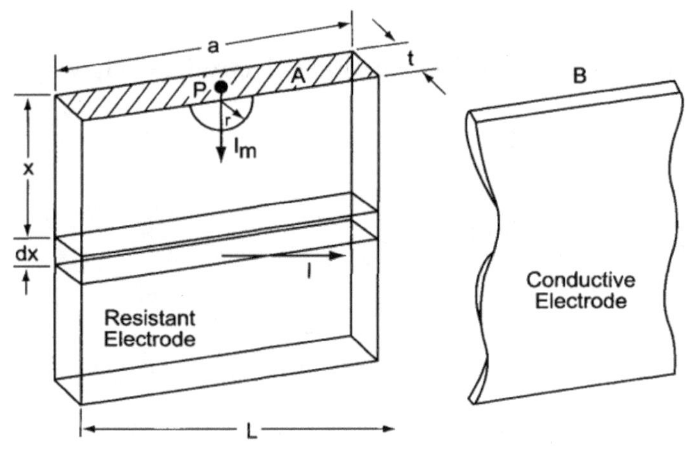

Figure 6.12. *Representation of a resistant electrode in parallel with a perfectly conductive electrode.*

Source: Janet L. White, Creative Commons License.

The Tafel equation is utilized to represent the correlation between the overvoltage and current density, denoted as i.

$$\eta = \frac{RT}{\beta F}\log\left(\frac{i}{i_o}\right)$$

The exchange current density and transfer coefficient is denoted by the symbols."i_o" and "β," respectively. The aforementioned equation may be regarded as linear within a restricted interval of current density, namely:

$$\eta = \frac{RT}{\beta F} \log\left(\frac{i}{i_o}\right)$$

where;

$$i = -\frac{di_m}{dx} = \frac{V - \eta}{R_s}$$

where; $R_s = r L$

$$-\frac{di_m}{dx} = \frac{V - \eta_0}{R_s[1 + (k/R_s)]}$$

Since,

$$-\frac{dV}{dx} = R_m i_m$$

where; is the electric resistance of the electrode, we have:

$$\frac{d^2 i_m}{dx^2} = \left[\frac{R_m}{R_s[1 + (k/R_s)]}\right] i_m$$

Solving above equation:

$$i_m(x) = i_m(0)e^{-\lambda_1 x}$$

$$\lambda_1^2 = \frac{R_m}{R_s[1 + (k/R_s)]} = \frac{R_m}{\rho L[1 + (k/\rho L)]}$$

$$\frac{i}{i(0)} = e^{-\lambda_1 x}$$

for $k>0$.

It follows that will be l when is small or when the polarization resistance is large. The Wagner number, denoted as k/L, holds significant importance in evaluating the impact of polarization on the distribution of current. It should be noted that in cases where the electrode kinetics exhibit sluggish behavior, a uniform distribution of current is observed irrespective of the electrode's shape (Matlosz et al., 1987).

Wagner conducted a study on the effects of overvoltage on the distribution of current in a finite plate electrode. The study was based on the premise of a linear relationship between the current density and overvoltage.

$$V_c = E + \eta_0 + \frac{k}{\rho}\left(\frac{dV}{dn}\right)$$

The determination of the cell voltage's position is contingent upon the thermodynamic decomposition voltage, symbolized as E, and the polarity of the current, denoted by the variable n.

The Laplace equation's solution was derived through the use of an analytical expression, which entails a Fredholm integral of the second kind. Figure 6.13 provides a visual representation of the aforementioned integral. The results suggest that an elevation in the value of k/results in a greater homogeneity in the dispersion of the electrical current (Dukovic, 1990).

The mathematical representation of the Wagner number can be achieved by employing the Tafel slope and the exchange current density, as illustrated below. Assuming a negligible reverse reaction rate, the rate equation can be expressed as follows:

$$i = i_o \exp(\eta/f)$$

$$f = \frac{RT}{\beta F}$$

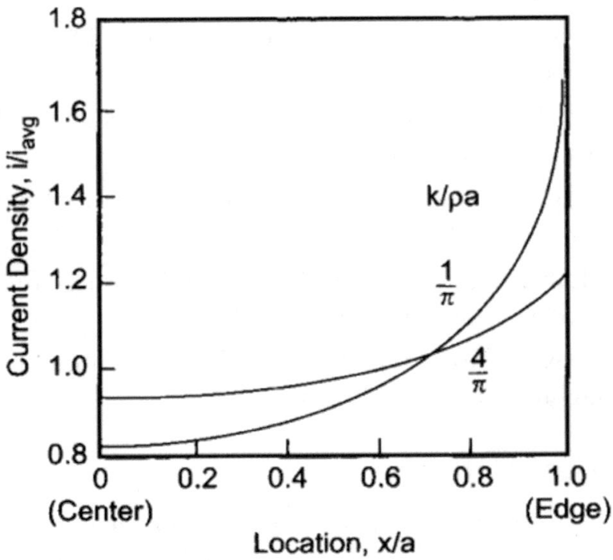

Figure 6.13. *Effect of polarization on the current distribution on the electrode.*

Source: Brent Hardman, Creative Commons License.

It follows from the equation above that:

$$\frac{d\eta}{di} = k = \frac{f}{i_o \exp(\eta/f)} = \frac{f}{i}$$

Thus,

$$W_A = \frac{f}{i\rho L}$$

6.3.3. Tertiary Current Distribution

The distribution of tertiary current is subject to the influence of multiple factors, including but not limited to cell geometry, kinetics, mass transfer, and electrode. Efficient computational processing is required for the calculation of complex tertiary current distribution solutions. The process of solving equations simultaneously is characterized by a significant level of intricacy:

- The equation governing the momentum of a velocity vector component.
- The concentration profile in the context of connective diffusion can be described by an equation.
- The acquisition of the current density profile is imperative as it remains undisclosed.
- The voltage balance equation is a fundamental concept in electrical engineering.
- The task at hand is to finalize the exponential electrochemical rate equation.
- The Laplace equation is utilized in bulk through the utilization of electrode surface current distribution.
- The expression for logarithmic concentration overvoltage (Sautebin & Landolt, 1982).

6.3.4. Numerical Methods

The existence of intricate geometries and non-linear boundary conditions poses a hindrance to the development of analytical solutions for current distribution problems. Numerous noteworthy research works have computed and arranged the fundamental concurrent dispersion patterns for diverse geometries. Prentice and Tobias conducted a comprehensive analysis of various methodologies employed in resolving numerical distribution problems. The current distribution profiles are determined by employing both finite difference and finite element methods.

Computational resources and commercial solvers are available for addressing boundary value problems.

- ***Examples:*** The preceding text provides a succinct summary of the fundamental principles that form the basis for determining the distribution of current and potential between electrodes within an electrolytic medium. The primary objective of this document is to furnish the reader with a comprehensive comprehension of the fundamental principles and to steer them toward pertinent scholarly works pertaining to the topic (Sharma et al., 2020).

6.3.4.1. Vertical Electrolyzers

The diaphragm and membrane cells exhibit a vertical orientation. The bipolar electrolyzer operates by receiving energy from two opposing ends within the cell stack, resulting in the generation of a perpendicular current to the electrode. The vertical electrolyzer of the tank type experiences ohmic voltage dips due to the energy input received from either the top or bottom electrode.

The present investigation analyzes the behavior of a flat-plate electrode operating at a voltage of zero within a cell that is arranged parallel to a counter electrode characterized by perfect conductivity. The resistive electrode is subjected to current input from the upper region. The decrease in electrolytic current across the working surface is attributed to the presence of ohmic resistance in the electrode. The circuit that is analogous provides the current density, denoted as *i*(x), at a specific distance x from the source of current (Hine & Murakami, 1980).

$$i(x) = i_o \exp(-\lambda x)$$

$$\lambda^2 = \frac{R_m}{R_s} = \frac{R_m}{\rho L}$$

where; $i(x)$ is the current density with unit as A ; is the current density at $x = 0$, with unit as A ; is the electric resistance of electrode, with unit as Ω; is the electric resistance of electrolyte, with unit as Ω; is the electric resistivity of electrolyte, with unit as Ωcm; and L is the inter electrode gap, with unit as cm.

At the top of the cell, gaseous components separate from the solution, while bubble-free solution sinks to the bottom. Internal flow decreases the solution voltage and lowers the gas void percentage. The design of downcomer solution-based vertical cells with gas-evolving electrodes is crucial.

6.3.4.2. Relative Disposition of Anodes and Cathodes in Diaphragm Cells

It is recommended that diaphragm cells possess equivalent heights for both the anode and cathode. In the event of failure, it is recommended that the anode be of greater height than the cathode. The principal current distribution of L-shaped cathode geometry was invented by Klingert. The simplified cathode top is depicted in Figure 6.14. The variable α denotes the height of the brine solution above the cathode, while β represents the distance between the anode and cathode. The B position denotes the cathode edge (Grotheer et al., 2006).

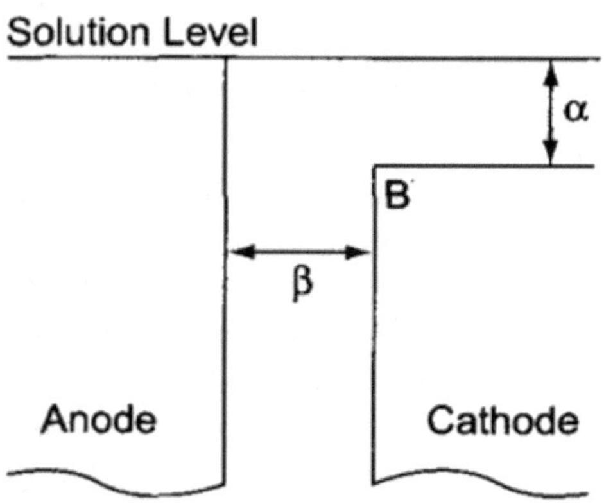

Figure 6.14. *Schematic of electrodes of different heights immersed in a solution.*

Source: Tilak V. Bommaraju, Creative Commons License.

$$\frac{dz}{dt} = C_1 \frac{\sqrt{t^2 - b^2}}{t^2 - a^2}$$

The potential field in the vicinity of the cathode edge B can be analyzed through the utilization of conformal transformation.

The transformation of the upper half of the i-plane into the z-plane is achieved through conformal mapping using the Schwarz Christoffel method, as follows:

Substituting:

T=snu

above with a = I and b = l/k and integrating it results in:

$$z = C_1 \int \frac{\sqrt{sn^2u - 1/k^2}}{sn^2 - 1} cnu \cdot dn\, u\, du + C_2$$

$$= jC_1 \int dn^2u \cdot du + C_2 = -jC_1 E(u) + C_2$$

$$u = 0 \quad \text{at } z = 0 \text{ and hence } C_2 = 0$$

$$z = -jC_1 E(u)$$

Since $z = (\alpha + j\beta$ and $u = \mathrm{K} + \mathrm{K'}$ at point B, we have:

$$C_1 = \frac{\alpha + j\beta}{K' - E' - jE}$$

Substituting in the equation above results in the following:

$$z = \frac{E(u)}{(K' - E')^2 + E^2}[\alpha(E - E' + K') + \beta(K' - E') + j\beta E]$$

Cathode surface current density is:

$$\frac{i}{i_0} = \frac{\partial u}{\partial z}$$

The distribution profile of the cathode finger is defined by the aforementioned relationships. The vertical axis of the diagram represents the local current density observed on the top surface of the cathode finger, whereas the horizontal axis denotes the distance measured from the edge. The distribution is impacted by the α/β ratio. The occurrence of diaphragm pluggage results in a reduction of the liquid level (e) in the catholyte, leading to a lack of cathodic protection on the upper surface of the cathode. This, in turn, exacerbates the corrosion process. The length of the shielded area xL of the cathode increases in instances of high liquid levels. Optimal cathodic protection current density can be achieved by maintaining high levels within the prescribed design and operational limitations, as it has been observed that the current density for cathodic protection tends to increase with rising temperatures and caustic concentrations (Florkiewicz, 2008).

The present study involves the computation of distances originating from the acute apex situated atop the digit while considering cathodic protection current of 10% as a precautionary measure. A protective ratio of α/β 0.25

can provide adequate protection to the weld when the distance between the welds and the sharp corner is 0.6 cm, and the value of β ranges from 1.5–1.8 cm. Cathodic protection is achieved for welds when the anode is positioned at a height of 6.35 mm above the top of the cathode finger (Shim et al., 2015).

6.3.4.3. Cathode Corrosion

The MDC-29 diaphragm cells are subject to cathode corrosion in the vicinity of the screen located at the interface between the cell liquor and hydrogen. The deposition of high-quality diaphragm is impeded by corrosion. The finite difference method was utilized to compute the primary current distributional profiles. Case A exhibits a cathodic protection current of 0.93 kA and an average cathode current density of 2.17 kA. As a result, there is a possibility of corrosion occurring at the corners of the cathode screen. Cell designs B and C exhibit efficient current distribution and is resistant to corrosion. The extension of anodes is a viable solution to mitigate the effects of corrosion. The application of cathodic protection involves the augmentation of electrical current in regions that are susceptible to corrosion (Hersbach & Koper, 2021).

6.4. FLUID PROCESSING

6.4.1. Fluid Dynamics

6.4.1.1. Bernoulli's Theorem

The diagram depicted in Figure 6.15 illustrates a tube in a state of equilibrium. During a specific time, interval dt, the mass of fluid entering the tube from the left and exiting from the right-hand side is indicated as $r_1 A_1 u_1 dt$ and $r_2 A_2 u_2 dt$, respectively. In this context, the symbol "p" denotes the density of the fluid, "A" signifies the cross-sectional area of the tube, and "u" represents the velocity of the fluid flow (Schar, 1993). It should be noted that the volumes of the fluid sections are equal, ensuring an equilibrium state:

$$r_1 A_1 u_1 = r_2 A_2 u_2$$

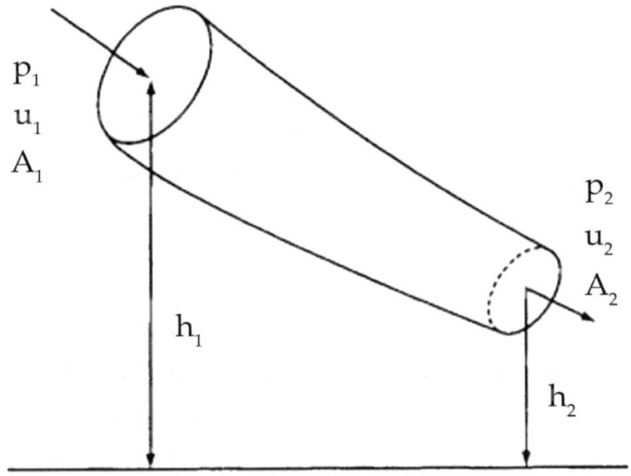

Figure 6.15. *Schematic of liquid flow through a pipe.*

Source: Tilak V. Bommaraju, Creative Commons License.

Also,

$$A_1u_1 = A_2u_2$$

The equations mentioned above are commonly referred to as the equations of continuity in academic literature.

The present discourse pertains to the phenomenon of gravity-driven flow of a fluid that exhibits incompressibility. The kinetic energy associated with the fluid motion over a given period of time. The expressions for potential energy and pressure energy are given by $(r\,A_1u_1dt)\,gh_1$ and pA_1u_1dt, respectively. Here, g represents the acceleration due to gravity, h denotes the elevation and represents the pressure. Given that the conservation of energy principle applies to the flow tube, the total energy within it remains constant (Pearson, 1925).

$$\frac{1}{2}(\rho A_1 u_1\,\mathrm{d}t)u_1{}^2 + (\rho A_1 u_1\,\mathrm{d}t)gh_1 + p_1 A_1 u_1\,\mathrm{d}t$$
$$= \frac{1}{2}(\rho A_2 u_2\,\mathrm{d}t)u_2{}^2 + (\rho A_2 u_2\,\mathrm{d}t)gh_2 + p_2 A_2 u_2\,\mathrm{d}t$$

The above equations lead to the following:

$$\frac{1}{2}\rho u_1^2 + \rho g h_1 + p_1 = \frac{1}{2}\rho u_2^2 + \rho g h_2 + p_2$$

or

$$\frac{1}{2}u^2/g + p/\rho g + h = \text{constant}$$

In order to express pressure, denoted as p, in engineering units, it is necessary to incorporate the typical acceleration due to gravity into the aforementioned equation.

$$\frac{1}{2}u^2/g + pg_c/\rho g + h = \text{constant}$$

The three components on the left-hand side of the aforementioned equations possess the physical dimensions of height. As such, they are individually referred to as the velocity head, the pressure head, and the potential head. Stated differently, the initial and subsequent terms are indicative of the fluid elevations necessary to sustain a flow rate of u at the static pressure level of p (Miki, 1978).

6.4.1.2. Flow Pattern

When a fluid comes into contact with a wall, it experiences a state of rest due to the presence of frictional forces. The aforementioned quantities exhibit direct proportionality with the viscosity of the fluid. Whilst the velocity may exhibit a gradual increase as the distance from the wall increases, it is important to note that an internal frictional force persists. When fluid is permitted to flow along the x-axis, it generates a velocity gradient in the y-axis direction. The force caused by friction, *f*, is expressed as:

$$f = \mu A(\mathrm{d}u/\mathrm{d}y)$$

The symbol μ denotes the viscosity. The flow of an incompressible fluid in a straight pipe with a small concentric element of diameter 2r and length L, as depicted in Figure 6.16, may be contemplated at a relatively low-velocity u (Cheng et al., 2008).

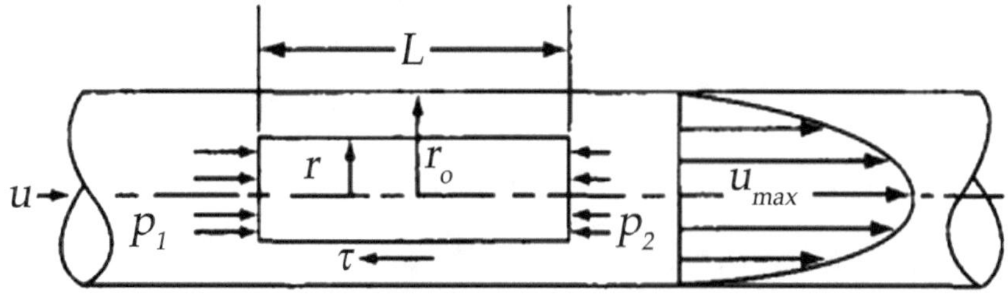

Figure 6.16. *Schematic of laminar flow in a pipe (Poiseulle flow).*

Source: Fumio Hine, Creative Commons License.

The friction terms on the inside surface and the outside surface of this element are:

$$f = 2\pi r L\mu(du/dr)$$

$$f + df = 2\pi r L\mu(du/dr) + d(2\pi r L\mu\, du/dr)$$

The force on a small ring, *dr,* caused by the pressure drop $(p_1 - p_2)$ is $2\pi r dr\ p_1 - p_2)$, and is equal to *df,* that is,

$$d(2\pi r L\mu\, du/dr) = 2\pi r(p_1 - p_2)\, dr$$

or

$$L\mu\, d(r\, du/dr) = (p_1 - p_2) r\, dr$$

Integrating this twice, we have:

$$u = -(p_1 - p_2)r^2/4L\mu + \text{constant}$$

Since the flow velocity at the wall is zero:

$$u = (p_1 - p_2)\left(r_0^2 - r^2\right)/4L\mu$$

Hence, the maximum flow velocity at the center of the pipe is given by the equation below:

$$u_{\max} = (p_1 - p_2) r_0^2/4L\mu$$

The volume of flow, in unit time, can now be obtained by integration as:

$$V = \int_0^{r_0} 2\pi r \cdot u \cdot dr = \frac{\pi(p_1 - p_2)}{8L\mu} r_0^4$$

Also, the average flow velocity u_{avg}, can be obtained by dividing the above equation by the sectional area, πr_o^2, as:

$$u_{\text{avg}} = (p_1 - p_2) D^2/32L\mu = u_{\max}/2$$

The variable D represents the diameter of the pipe. The aforementioned equation demonstrates that the decrease in pressure in the laminar flow regime is directly proportional to the velocity of the flow. Additionally, it has been established that the viscosity of a fluid is responsible for generating laminar flow at velocities that are relatively low. Upon exceeding a specific threshold, the magnitude of the inertial force becomes significant in comparison to the viscosity. The subject matter being addressed pertains to the widely recognized occurrence of turbulent flow. The thorough analysis of turbulent flow poses considerable difficulty. It is recognized that the shearing stress demonstrates a linear relationship with the square of the velocity gradient that is orthogonal to the solid/fluid interface (Kreutzer et al., 2005).

$$f = \rho L_{\mathrm{m}}^{2}(\mathrm{d}u/\mathrm{d}y)^{2}$$

where; is the length of mixing, which is analogous to the mean free path in the kinetic theory of gases.

The pressure drop is also proportional to the average flow velocity, as described by the equation below, which is known as the Fanning equation or the Darcy formula.

$$p_1 - p_2 = 4F_f (L/D)(u_{avg}^2/2)$$

in the equation above is termed Fanning's friction factor. We may now extend the above equation to the region of laminar flow. This results in the following:

$$F_f = 16/\mathrm{Re}$$

$$\ddot{u} = Du_{\pm} \; r \; m$$

The evidence suggests that there exists an inverse relationship between the variables of interest, denoted as and, specifically, when the value of is less than 2,300. The aforementioned equation exhibits a growing deviation at high Reynolds numbers. The upper limit or critical value for laminar flow is estimated to be approximately 2,006–2,300. Turbulent flow is initiated at Reynolds numbers exceeding 4,000. Within the interval of 2,300–4,000, the flow is in a state of transition. The surface morphology of the solid in contact with the fluid, specifically the pipe, has a significant impact on the friction factor in the turbulent flow regime (Zhang et al., 2012).

Numerous mathematical formulations exist for modeling flow in a pipe under turbulent flow conditions, in addition to several empirical equations that establish a relationship between certain variables.

6.4.1.3. Pressure Drop Calculations

Circular piping is a widely used method for fluid transportation in industrial processes, which relies on pressure differential. The circular arrangement offers both enhanced structural stability and an ideal cross-sectional area in proportion to the wall surface per unit (Taitel & Barnea, 1990).

The calculation of pressure drop is a critical component of pipeline design and optimization, as well as the selection of suitable pump and compressor sizes for fluid conveyance. This section presents a discourse on equations commonly employed in the calculation of pressure drops (Russo et al., 2016).

6.4.1.4. Non-Compressible Fluid Flow

The utilization of the Darcy formula is a viable approach for evaluating the movement of incompressible fluids. This principle holds true in most cases, with the exception of scenarios where the reduction in liquid pressure is substantial enough to cause the outlet pressure to drop below the liquid's vapor pressure. To address this situation, individuals may choose to alter the dimensions of the pipe, decrease the rate of flow, or create a counteracting pressure (Kishna & Vajravelu, 2022).

$$u_{avg} = Q / A$$

The physical property data is utilized to acquire information regarding the density and viscosity. The Reynolds number serves as an indicator of whether the flow regime is laminar or turbulent.

One can compute the loss of fluid in feet by means of friction, denoted as.

$$H_L = f(L / D)(u^2 / 2gc)$$

6.4.1.5. Compressible Fluid Flow

In the realm of fluid mechanics, a fluid is deemed non-compressible when alterations in pressure and temperature are of a magnitude that does not result in consequential modifications to its density. Flow is deemed compressible when the density undergoes a variation exceeding 5 or 10%. The expression for the pressure drops under the specified conditions is provided (Taylor & Sharman, 1928).

$$p_1^2 - p_2^2 = G^2(RT/M_W)[(4fL/D_H) + 2\ln(p_1/p_2)]$$

where G is a mass flow rate per unit area, R is the ideal gas constant, Mw is the molecular weight of the fluid, L is the equivalent pipe length, Du is the hydraulic diameter, and pi and p2 are the pressures at the beginning and end of the pipe, respectively. Friction loss computations are applicable to pipes that are uniform and straight. Valves and fittings in pipelines have the potential to cause disturbances in flow patterns, generate turbulence, and elevate pressure losses. The losses incurred during fluid flow must be considered in the calculations. The assignment of a linear conduit segment to every valve and fitting is a commonly employed technique. Equivalent pipe sizes are utilized in textbooks and handbooks to present data. The conventional measurement for the length of elbows is 30 times the diameter of the pipe (Lamont et al., 1997). The pipe in question has a diameter of 100 millimeters and a linear distance of 3 meters without any bends or curves. The determination of the comparable length of a pipe system involves the inclusion of the aggregate lengths of all valves and fittings up to the final linear segment. The symbol "L" denotes a length that can be compared to another length. This straightforward technique can be employed for manual computations. Theoretically speaking, it lacks validity. The determination of L/D values as a function of D is necessary for the geometric analysis of fittings with varying sizes. Resistance coefficients are employed in lieu of other measures (Sun & Agafonov, 2012).

6.4.1.6. Hydraulic Diameter

The flow in pipes having a circular cross-section of in diameter was described in earlier sections. Diverse geometries of piping are utilized in chemical processing facilities. The utilization of the Fanning equation is applicable in such scenarios, whereby the variable D is substituted with the hydraulic diameter or its equivalent.

$$D_{\mathrm{H}} = 4A/L_{\mathrm{W}} = 4R_{\mathrm{H}}$$

where; is substituted for *D*. In the equation above, *A* refers to the sectional area of the channel under consideration, is the length of the solid/fluid interface or wetted perimeter, and EH is the hydraulic radius. The values of hydraulic radius for various cases are discussed in the section on fluid dynamics (Lewis, 1992).

- **Pressure Drops Caused by Geometry:** The dissipation of hydraulic energy occurs due to irregularities, such as bends and elbows, present in a pipeline. Abrupt variations in the cross-sectional area of fluid-carrying conduits, whether constrictions or expansions, result in losses at the inlet and outlet sections. Abrupt variations in the cross-sectional area also result in an elevation of the pressure drop (Dietrich et al., 2009).

6.4.2. Removal of Solids

The elimination of particulate matter from gaseous streams holds significant importance in various industrial operations and serves as a crucial measure for safeguarding the environment against the discharge of stack gases and similar emissions. The matter at hand holds a higher degree of significance, and therefore, this chapter does not delve into the aforementioned topic. The separation of particulate matter from fluids is a crucial aspect of the handling of corrosive substances and is especially significant in the context of the facility responsible for the treatment of saline solutions. The involved unit operations comprise gravitational sedimentation, filtering, and centrifugation. The process of gravity sedimentation is employed to eliminate solid particles from brine that has undergone chemical treatment (Paterson & Yao, 1990). The formation of solids occurs through the introduction of chemical agents that induce the precipitation of ionic impurities from the solution. As the separation process remains incomplete, it becomes imperative to supplement the sedimentation procedure with brine filtration. Polishing filtration holds significant importance on the caustic end, particularly in the context of mercury-cell plants. The elimination of salt precipitates from the caustic product in evaporators is a requisite procedure in diaphragm-cell facilities. The majority of the salt is extracted through a multi-step centrifugation process. The removal process exhibits incompleteness, thereby necessitating the filtration of the liquor derived from the centrifuges. Typically, the sedimentation process of processed brine results in the formation of a sludge that exits the settling equipment as a diluted mixture. Several categories of filters are employed in numerous industrial facilities to enhance the concentration of solids prior to their disposal (Buttress et al., 2016).

The elimination of mists from chlorine gas is a distinct instance of filtration. Two distinct applications warrant consideration in this context. The first pertains to chlorine derived from the cooling system, which may potentially retain minute particles entrained from the cells. The second application involves chlorine originating from the drying towers, which is accompanied by a mist of sulfuric acid. The equipment utilized in such instances typically depends on the Brownian motion of particles within the separation components. The operational process of this mechanism exhibits notable distinctions from that of traditional filtration (Sarika et al., 2005).

The process of sedimentation is predicated on the force of gravity to effectuate the separation of heavier particulate matter from the comparatively less dense fluid phase. The phenomenon of settling is impeded by the counteracting forces of friction that exist between substances in solid and liquid states. The properties of the particles generated in the process of precipitation play a crucial role in the formation of sediments, and this constitutes the focal point of investigation. The process of sedimentation analysis can be simplified if the solid particles are assumed to be equivalent spheres and their size is measured using the Sauter mean diameter (Haas, 1998). However, this method is only effective when the concentration of the particles is relatively low. The complexity of the separation mechanism increases as the concentration reaches a level that permits inter-particle interactions. Subsequently, it becomes imperative to employ empirical techniques for computation. Several techniques employed are delineated (Yang et al., 2018).

The complexity of analyzing the filtration process arises from the continuous alteration in flow resistance due to the gradual accumulation of solid matter on the filter medium.

The analysis of flow through packed beds of solids typically involves an examination of key features such as the bed's porosity and the sphericity of its constituent particles. In addition, the study of filters can be facilitated by an assessment of how the deposition of precipitated solids alters these characteristics. The remaining solids in the other filtration processes accumulate as a layer on the filter medium, commonly referred to as a cake. The resistances exhibited by both the filter cake and medium are subsequently combined. When the resistivity of the cake remains constant during operation, the specific resistance exhibits a linear increase with the quantity of solid deposited per unit thickness. Subsequently, it becomes feasible to obtain analytical solutions for the filtration rate. In the scenario where the rate remains constant, the pressure differential experienced can be mathematically represented as a function of time (Kumari et al., 2015).

The resistivity of the cake frequently experiences an increase over time. As a consequence of the filtration pressure or the fluid friction, the particles that have been filtered undergo compaction and subsequently diminish the available flow area. Furthermore, the text elucidates the empirical techniques employed in characterizing the compressibility of the cake. The compressibility of particles resulting from precipitation from brine is a noteworthy aspect. Simultaneously, there exists a proclivity for them to become trapped within the interstices of the filtration material. The filtration process of these particles may pose significant challenges (Chu

& Rochelle, 1989). Typically, a filter aid is employed as the conventional approach to address this issue. Prior to the introduction of brine, the filter aid is administered onto the filter medium's surface. This is the preliminary coating process. Filter aids are chosen based on their favorable attributes, and the pre-coating substance safeguards the pores against infiltration by filtered solids while simultaneously providing minimal impedance to the flow. The process of precipitate compression persists, prompting the addition of filter aid, also known as "admix," in minimal amounts to the brine to enhance the cake's properties (Antonioni et al., 2016).

Centrifuges are commonly utilized as proficient filters, whereby the separating force is amplified through the operation of the apparatus. The primary utilization within the chlor-alkali sector pertains to the process of diaphragm-cell caustic evaporation. Approximately 3% of the evaporator product is constituted by the salt that was precipitated during said operation. The majority of this substance is separated through the process of centrifugation (Balasubramanian & Balasubramaniam, 2003).

6.4.3. Compression and Liquefaction

6.4.3.1. Compression

The equations governing the behavior of an ideal gas in a state of equilibrium are presented. Several of these variables hold significance in the examination of gas compression.

$$dQ = dU + dW$$

$$dW = -p\,dV$$

$$c_p = dH/dT$$

$$c_v = dU/dT$$

$$c_p - c_v = R$$

We also use the ratio.

$$k = c_p / c_v$$

Combining the above equations:

$$dQ = c_v\,dT + p\,dV$$

in an isothermal process:

$$dQ = dW = p\,dV$$

For an ideal gas pV = RT. Then,

$$dQ = RT\,dV/V$$

integrating from the starting volume,

$W=Q=RT\ln(V/V_1)$

In the adiabatic case, Q = 0.

$$\ln(T/T_1) = (R/c_v)\ln(V_1/V)$$

$R/c_v = k - 1$. Now,

$$T/T_1 = (V/V_1)^{1-k}$$

The aforementioned equation enables the computation of the rise in gas temperature resulting from adiabatic compression, contingent upon the level of compression and specific heat ratios. Utilizing the ideal gas equation once more, it can be inferred that we possess.

$$pV^k = p_1 V_1^k = \text{constant}$$

As work is a non-point function, the outcome is contingent on the trajectory taken during the course of the process. In the scenario where adiabatic conditions are present, the amount of work that is required to be provided across an equivalent range of pressure remains constant.

$$W = p_1 V_1 k/(k-1)[(p_2/p_1)^{[(k-1)/k]} - 1]$$

In the case of non-ideal gases, it is necessary to multiply this value by the compressibility factor, Z. Regarding chlorine, the omission of Z yields a negligible error, which is also a conservative estimate. Chlorine's critical pressure of 77.2 bars is seldom surpassed during its production process compression, which typically ranges from 0.01–0.1 reduced pressures. Under the specified inlet conditions, the reduced pressure exhibits a considerable degree of diminution, resulting in a Z factor of approximately 0.98 to 0.99. At low temperatures, Z is expected to be around 0.95, while at typical compressor outlet temperatures, it is anticipated to be around 0.98 when the gas undergoes compression (Kim et al., 2004).

Under standard inlet conditions, the value of k typically ranges between 1.32 and 1.35. The aforementioned value is marginally inferior to the specific heat ratio of air. Charts that display the energy needed for air compression offer a conservative margin of safety when utilized for approximating the compression energy necessary for chlorine. The expression (k l)/k, frequently encountered in compression computations, is estimated to be approximately 0.25. Upon assigning the compression ratio as "r," the aforementioned equation is transformed (Quirk et al., 2004).

$$W = 4RT_1(r^{0.25} - 1)$$

The specific heat ratio's actual worth is contingent upon the prevailing operational circumstances. The specific heat ratio experiences a slight decrease from the feed to the discharge across the entire range of operation of a chlorine compressor.

Of greater importance is the observation that an actual compression process deviates from being strictly adiabatic (Luty & Eckhardt, 1995). It follows a polytropic path in which:

$$pV^n = \text{constant}$$

The value of the exponent n is determined through empirical observation and experimentation. In all instances, the level of power consumption is higher compared to the adiabatic scenario. The exponent "ii" is invariably greater than "k," and there exists a correlation between the two through the polytropic efficiency.

$$n = k\eta_p/[k\eta_p - (k - 1)]$$

The polytropic efficiency of 75% yields a value of 3k/(4 k), which approximates 1.5. The empirical data suggests that the optimal values of n for chlorine and hydrogen are approximately 1.45 (Hicks et al., 2009).

6.4.3.2. Liquefaction

The process of liquefaction necessitates the addition of two forms of energy: the sensible heat that is extracted from the gas during the cooling process until it reaches the temperature required for liquefaction and the latent heat of condensation:

$$Q = qv + \lambda$$

The equation for determining the energy removed in the liquefaction of a specific amount of gas is represented by Q, while the sensible heat required to decrease the temperature to the point of liquefaction is represented by λ.

It refers to the amount of latent heat that is extracted at a constant temperature in order to condense a gas.

In the event that the liquid is subcooled, an additional quantity of sensible heat is involved. The quantity representing the amount of heat energy absorbed or released by a gas while its pressure remains constant, commonly referred to as the heat capacity of the gas at constant pressure, can be calculated (Poulos et al., 1985).

$$\mathrm{d}q_{\mathrm{L}} = c_{\mathrm{p}}\,\mathrm{d}t$$

where; c_p is the specific heat of the gas at constant pressure.

q_v, in the above equation, has the potential to be assessed through analytical means, provided that a mathematical representation of the particular heat is obtainable. Numerous publications have featured a variety of equations, with two particular formulas being commonly utilized.

$$c_{\mathrm{p}} = a + bT + cT^2$$

$$c_{\mathrm{p}} = a + bT + c/T^2$$

The equations presented above, which may incorporate varying values for the constants, occasionally employ Celsius temperature units. In certain instances, the inclusion of a cubic term may be deemed appropriate for an extension. The utilization of absolute temperature is required for the final term in Eqn. (46) (Seed & Idriss, 1971).

At identical conditions, the latent heat of condensation is equivalent to the latent heat of vaporization. Thermodynamic equations are commonly derived with respect to the process of vaporization, utilizing the latent heat of vaporization, denoted as λ. The latent heat plays a crucial role in the initiation of vapor formation, as it serves as a type of activation energy (Williams & Larson, 2003). Additionally, it determines the gradient of the vapor pressure curve. The Clausius-Clapeyron equation is at our disposal.

$$\mathrm{d}P^{\mathrm{o}}/\mathrm{d}T = \lambda/T(V - v)$$

where; $P°$ is the vapor pressure; V is the molar volume of gas; and r is the molar volume of liquid.

when; V >> v, we have, with ideal gas behavior:

$$\mathrm{d}P^{\mathrm{o}}/P = (\lambda/R)\,\mathrm{d}T/T^2$$

Integration yields:

$$\ln(P_2/P_1) = -(\lambda/R)(1/T_1 - 1/T_2)$$

The heat of the vaporization process of a substance can be determined through the utilization of the aforementioned equation based on its vapor pressure curve. In practical applications, the gas supplied for the liquefaction process typically contains a certain amount of non-condensable gas (Uzuoka et al., 1998).

$$q_V = q_{Cl} + q_I$$

where; is the sensible heat removed from chlorine gas; is the sensible heat removed from non-condensable gases.

The influence of the liquefaction temperature on the cost of energy removal is a critical factor to be taken into account during the development of liquefaction systems. The observation has been made that a decrease in liquefaction temperature leads to a more efficient liquefaction process. However, this is accompanied by a notable increase in the cost of energy removal. The aforementioned information offers a justification for this occurrence. The main cost pertains to the supply of energy required to operate the refrigeration apparatus. Let us consider the thermodynamic cycle of the refrigerant (Okamura & Soga, 2006).

6.5. SUMMARY

Chemical engineering principles play a crucial role in the chlor-alkali industry. The production of chlorine and caustic soda in the chlor-alkali industry relies on the electrolysis of brine. In this process, a direct current is passed through a brine solution, leading to the electrochemical reactions that produce chlorine gas and sodium hydroxide. Chemical engineers in the chlor-alkali industry are responsible for ensuring safe operating conditions. This includes managing and mitigating potential risks associated with handling and storing of chlorine gas which is toxic and hazardous.

REFERENCES

1. Alvarez, A. J., & Myerson, A. S., (2010). Continuous plug flow crystallization of pharmaceutical compounds. *Crystal Growth and Design, 10*(5), 2219–2228.
2. Antonioni, G., Dal Pozzo, A., Guglielmi, D., Tugnoli, A., & Cozzani, V., (2016). Enhanced modelling of heterogeneous gas–solid reactions in acid gas removal dry processes. *Chemical Engineering Science, 148*, 140–154.
3. Balasubramanian, S., & Balasubramaniam, V. M., (2003). Compression heating influence of pressure transmitting fluids on bacteria inactivation during high pressure processing. *Food Research International, 36*(7), 661–668.
4. Bernardi, D., Pawlikowski, E., & Newman, J., (1985). A general energy balance for battery systems. *Journal of the Electrochemical Society, 132*(1), 5.
5. Borja, R. I., (2006). Condition for liquefaction instability in fluid-saturated granular soils. *Acta Geotechnica, 1*, 211–224.
6. Buttress, A. J., Binner, E., Yi, C., Palade, P., Robinson, J. P., & Kingman, S. W., (2016). Development and evaluation of a continuous microwave processing system for hydrocarbon removal from solids. *Chemical Engineering Journal, 283*, 215–222.
7. Cardella, U., Decker, L., & Klein, H., (2017). Economically viable large-scale hydrogen liquefaction. In: *IOP Conference Series: Materials Science and Engineering* (Vol. 171, No. 1, p. 012013). IOP Publishing.
8. Chambré, P. L., & Acrivos, A., (1956). On chemical surface reactions in laminar boundary layer flows. *Journal of Applied Physics, 27*(11), 1322–1328.
9. Cheng, L., Ribatski, G., & Thome, J. R., (2008). Two-phase flow patterns and flow-pattern maps: Fundamentals and applications. *Applied Mechanics Reviews, 61*(5).
10. Chi, J. W. H., (1965). Cooldown temperatures and cooldown time during mist flow. In: *Advances in Cryogenic Engineering: Proceedings of the 1964 Cryogenic Engineering Conference (Sections AL)* (pp. 330–340). Springer US.
11. Chu, P., & Rochelle, G. T., (1989). Removal of SO_2 and NOx from stack gas by reaction with calcium hydroxide solids. *Japca, 39*(2), 175–179.
12. Cripps, S. J., & Bergheim, A., (2000). Solids management and removal for intensive land-based aquaculture production systems. *Aquacultural Engineering, 22*(1–2), 33–56.

13. Crossley, S., Whatmore, R. W., Mathur, N. D., & Moya, X., (2021). Quasi-indirect measurement of electrocaloric temperature change in PbSc0. 5Ta0. 5O_3 via comparison of adiabatic and isothermal electrical polarization data. *APL Materials, 9*(1), 010701.

14. Datta, M., & Harris, D., (1997). Electrochemical micromachining: An environmentally friendly, high speed processing technology. *Electrochimica Acta, 42*(20–22), 3007–3013.

15. Dietrich, B., Schabel, W., Kind, M., & Martin, H., (2009). Pressure drops measurements of ceramic sponges—determining the hydraulic diameter. *Chemical Engineering Science, 64*(16), 3633–3640.

16. Dou, B., Wang, C., Song, Y., Chen, H., Jiang, B., Yang, M., & Xu, Y., (2016). Solid sorbents for *in-situ* CO_2 removal during sorption-enhanced steam reforming process: A review. *Renewable and Sustainable Energy Reviews, 53*, 536–546.

17. Dukovic, J. O., (1990). Computation of current distribution in electrodeposition: A review. *IBM Journal of Research and Development, 34*(5), 693–705.

18. Ensing, B., De Vivo, M., Liu, Z., Moore, P., & Klein, M. L., (2006). Metadynamics as a tool for exploring free energy landscapes of chemical reactions. *Accounts of Chemical Research, 39*(2), 73–81.

19. Fabre, J., & Liné, A., (1992). Modeling of two-phase slug flow. *Annual Review of Fluid Mechanics, 24*(1), 21–46.

20. Florkiewicz, T. F., (2008). Advanced diaphragm cell technology. *Modern Chlor-Alkali Technology, 8*, 165.

21. Fuladpanjeh-Hojaghan, B., Elsutohy, M. M., Kabanov, V., Heyne, B., Trifkovic, M., & Roberts, E. P., (2019). In-operando mapping of pH distribution in electrochemical processes. *Angewandte Chemie, 131*(47), 16971–16975.

22. Garratt, J. R., (1990). The internal boundary layer: A review. *Boundary-Layer Meteorology, 50*, 171–203.

23. Georgiadou, M., (2003). Modeling current density distribution in electrochemical systems. *Electrochimica Acta, 48*(27), 4089–4095.

24. Grotheer, M., Alkire, R. C., Varjian, R., Srinivasan, V., & Weidner, J. W., (2006). Industrial electrolysis and electrochemical engineering. *The Electrochemical Society Interface, 15*(1), 52.

25. Guoxing, C., Enquan, Z., Zhihua, W., Binghui, W., & Xiaojun, L., (2016). Experimental investigation on fluid characteristics of medium dense saturated fine sand in pre-and post-liquefaction. *Bulletin of Earthquake*

Engineering, 14, 2185–2212.

26. Haas, P. A., (1993). A review of information on ferrocyanide solids for removal of cesium from solutions. *Separation Science and Technology, 28*(17–18), 2479–2506.
27. Hadush, S., Yashima, A., & Uzuoka, R., (2000). Importance of viscous fluid characteristics in liquefaction induced lateral spreading analysis. *Computers and Geotechnics, 27*(3), 199–224.
28. He, X., & Li, N., (2000). Lattice Boltzmann simulation of electrochemical systems. *Computer Physics Communications, 129*(1–3), 158–166.
29. Hersbach, T. J., & Koper, M. T., (2021). Cathodic corrosion: 21st century insights into a 19th century phenomenon. *Current Opinion in Electrochemistry, 26*, 100653.
30. Hicks, D. G., Boehly, T. R., Celliers, P. M., Eggert, J. H., Moon, S. J., Meyerhofer, D. D., & Collins, G. W., (2009). Laser-driven single shock compression of fluid deuterium from 45 to 220 GPa. *Physical Review B, 79*(1), 014112.
31. Hine, F., & Hine, F., (1985). Amalgam-type chlor-alkali industry. *Electrode Processes and Electrochemical Engineering*, 139–165.
32. Hine, F., & Murakami, K., (1980). Bubble effects on the solution IR drop in a vertical electrolyzer under free and forced convection. *Journal of the Electrochemical Society, 127*(2), 292.
33. Hine, F., (1985). Voltage balance and energy balance in an electrolytic cell. In: *Electrode Processes and Electrochemical Engineering* (pp. 73–107). Boston, MA: Springer US.
34. Hine, F., Tilak, B. V., & Viswanathan, K., (1986). Chemistry and chemical engineering in the chlor-alkali industry. *Modern Aspects of Electrochemistry, 18*, 249–302.
35. Hu, H., & Yang, W., (2008). Free energies of chemical reactions in solution and in enzymes with ab initio quantum mechanics/molecular mechanics methods. *Annu. Rev. Phys. Chem., 59*, 573–601.
36. Huggins, C., & Neal, W., (1942). Coagulation and liquefaction of semen: proteolytic enzymes and citrate in prostatic fluid. *The Journal of Experimental Medicine, 76*(6), 527–541.
37. Irandoust, S., & Andersson, B., (1988). Mass transfer and liquid-phase reactions in a segmented two-phase flow monolithic catalyst reactor. In: *Tenth International Symposium on Chemical Reaction Engineerin* (Vol. 1, No. 1, pp. 1983–1988).
38. Karlová, K., Strecka, J., & Madaras, T., (2016). *Isothermal Entropy Change*

and Adiabatic Change of Temperature of the Antiferromagnetic Spin-1/2 Ising Octahedron and Dodecahedron. arXiv preprint arXiv:1611.04296.

39. Kim, M. H., Pettersen, J., & Bullard, C. W., (2004). Fundamental process and system design issues in CO_2 vapor compression systems. *Progress in Energy and Combustion Science, 30*(2), 119–174.
40. Kreutzer, M. T., Kapteijn, F., Moulijn, J. A., & Heiszwolf, J. J., (2005). Multiphase monolith reactors: Chemical reaction engineering of segmented flow in microchannels. *Chemical Engineering Science, 60*(22), 5895–5916.
41. Krishna, M. V., & Vajravelu, K., (2022). Rotating MHD flow of second grade fluid through porous medium between two vertical plates with chemical reaction, radiation absorption, Hall, and ion slip impacts. *Biomass Conversion and Biorefinery*, 1–15.
42. Kugler, K., Ohs, B., Scholz, M., & Wessling, M., (2014). Towards a carbon independent and CO_2-free electrochemical membrane process for NH_3 synthesis. *Physical Chemistry Chemical Physics, 16*(13), 6129–6138.
43. Kumari, N., Bahadur, V., Hodes, M., Salamon, T., Kolodner, P., Lyons, A., & Garimella, S. V., (2010). Analysis of evaporating mist flow for enhanced convective heat transfer. *International Journal of Heat and Mass Transfer, 53*(15–16), 3346–3356.
44. Lakeland, D. L., Rechenmacher, A., & Ghanem, R., (2014). Towards a complete model of soil liquefaction: the importance of fluid flow and grain motion. *Proceedings of the Royal Society A: Mathematical, Physical, and Engineering Sciences, 470*(2165), 20130453.
45. Lamont, G. J., Vancouver, C. A., Voss, H. H., West Vancouver, C. A., Schwab, C., Daimler-Benz, A. G., & Daimler-Benz, A. G. D. E., (1997). *Embossed Fluid Flow Field Plate for Electrochemical Fuel Cells*.
46. Lewis, A. M., (1992). Measuring the hydraulic diameter of a pore or conduit. *American Journal of Botany, 79*(10), 1158–1161.
47. Liberman, M. A., Kiverin, A. D., & Ivanov, M. F., (2012). Regimes of chemical reaction waves initiated by nonuniform initial conditions for detailed chemical reaction models. *Physical Review E, 85*(5), 056312.
48. Liu, Z., Wainright, J. S., Huang, W., & Savinell, R. F., (2004). Positioning the reference electrode in proton exchange membrane fuel cells: calculations of primary and secondary current distribution. *Electrochimica Acta, 49*(6), 923–935.
49. Luty, T., & Eckhardt, C. J., (1995). General theoretical concepts for solid state reactions: quantitative formulation of the reaction cavity, steric compression, and reaction-induced stress using an elastic multipole

representation of chemical pressure. *Journal of the American Chemical Society, 117*(9), 2441–2452.

50. Macdonald, D. D., & Challingsworth, M. L., (1993). Thermodynamics of nickel–cadmium and nickel–hydrogen batteries. *Journal of The Electrochemical Society, 140*(3), 606.
51. Matlosz, M., Creton, C., Clerc, C., & Landolt, D., (1987). Secondary current distribution in a hull cell: Boundary element and finite element simulation and experimental verification. *Journal of the Electrochemical Society, 134*(12), 3015.
52. Miki, H., (1978). A relation between Bernoulli numbers. *Journal of Number Theory, 10*(3), 297–302.
53. Muller, R. P., & Warshel, A., (1995). Ab initio calculations of free energy barriers for chemical reactions in solution. *The Journal of Physical Chemistry, 99*(49), 17516–17524.
54. Newman, J. S., & Tobias, C. W., (1962). Theoretical analysis of current distribution in porous electrodes. *Journal of The Electrochemical Society, 109*(12), 1183.
55. Newman, J., & Tiedemann, W., (1993). Potential and current distribution in electrochemical cells: Interpretation of the half-cell voltage measurements as a function of reference-electrode location. *Journal of The Electrochemical Society, 140*(7), 1961.
56. Okamura, M., & Soga, Y., (2006). Effects of pore fluid compressibility on liquefaction resistance of partially saturated sand. *Soils and Foundations, 46*(5), 695–700.
57. Oloman, C., & Li, H., (2008). Electrochemical processing of carbon dioxide. *ChemSusChem: Chemistry & Sustainability Energy & Materials, 1*(5), 385–391.
58. Otashu, J. I., & Baldea, M., (2019). Demand response-oriented dynamic modeling and operational optimization of membrane-based chlor-alkali plants. *Computers and Chemical Engineering, 121*, 396–408.
59. Panigrahi, S., & Dubey, B. K., (2019). Electrochemical pretreatment of yard waste to improve biogas production: understanding the mechanism of delignification, and energy balance. *Bioresource Technology, 292*, 121958.
60. Paterson, M. S., & Yao, F. F., (1990). Efficient binary space partitions for hidden-surface removal and solid modeling. *Discrete and Computational Geometry, 5*, 485–503.
61. Pearson, K., (1925). James Bernoulli's theorem. *Biometrika*, 201–210.

62. Poulos, S. J., Castro, G., & France, J. W., (1985). Liquefaction evaluation procedure. *Journal of Geotechnical Engineering, 111*(6), 772–792.
63. Prentice, G. A., & Tobias, C. W., (1982). A survey of numerical methods and solutions for current distribution problems. *Journal of the Electrochemical Society, 129*(1), 72.
64. Quirk, R. A., France, R. M., Shakesheff, K. M., & Howdle, S. M., (2004). Supercritical fluid technologies and tissue engineering scaffolds. *Current Opinion in Solid State and Materials Science, 8*(3–4), 313–321.
65. Rose, B. A., Davis, G. J., & Ellingham, H. J. T., (1948). Studies in the thermodynamics of metallurgical reduction processes by electrochemical methods. *Discussions of the Faraday Society, 4*, 154–162.
66. Russo, R., Neff, L. P., Johnson, M. A., & Williams, T. K., (2016). Emerging endovascular therapies for non-compressible torso hemorrhage. *Shock (Augusta, Ga.), 46*(3), 12.
67. Sautebin, R., & Landolt, D., (1982). Anodic leveling under secondary and tertiary current distribution conditions. *Journal of The Electrochemical Society, 129*(5), 946.
68. Schär, C., (1993). A generalization of Bernoulli's theorem. *Journal of Atmospheric Sciences, 50*(10), 1437–1443.
69. Seed, H. B., & Idriss, I. M., (1971). Simplified procedure for evaluating soil liquefaction potential. *Journal of the Soil Mechanics and Foundations Division, 97*(9), 1249–1273.
70. Sharma, H., Patil, M., & Woolsey, C., (2020). A review of structure-preserving numerical methods for engineering applications. *Computer Methods in Applied Mechanics and Engineering, 366*, 113067.
71. Shim, J. H., Jeong, J. Y., & Park, J. Y., (2015). Effects of operating conditions on improving alkali and electrical efficiency in chlor-alkali diaphragm cell. *Int. J. Electrochem. Sci., 10*, 6338–6347.
72. Shkadov, V. Y., (1967). Wave flow regimes of a thin layer of viscous fluid subject to gravity. *Fluid Dynamics, 2*(1), 29–34.
73. Slater, C. S., & Hesketh, R. P., (2004). Incorporating green engineering into a material and energy balance course. *Chemical Engineering Education, 38*(1), 48–53.
74. Stewart, H. B., & Wendroff, B., (1984). Two-phase flow: Models and methods. *Journal of Computational Physics, 56*(3), 363–409.
75. Sun, Z., & Agafonov, V. M., (2012). Numerical modeling of a four-electrode electrochemical accelerometer based on natural convection: The Boussinesq flow model vs. the compressible flow models. *Russian*

Journal of Electrochemistry, 48(8), 835–842.

76. Taitel, Y., & Barnea, D., (1990). A consistent approach for calculating pressure drop in inclined slug flow. *Chemical Engineering Science, 45*(5), 1199–1206.
77. Taitel, Y., & Barnea, D., (1990). Two-phase slug flow. In: *Advances in Heat Transfer* (Vol. 20, pp. 83–132). Elsevier.
78. Tishin, A. M., & Spichkin, Y. I., (2002). Relationship of adiabatic, isothermal, and field constant changes of a magnetic entropy. In: *AIP Conference Proceedings* (Vol. 614, No. 1, pp. 27–34). American Institute of Physics.
79. Uzuoka, R., Yashima, A., Kawakami, T., & Konrad, J. M., (1998). Fluid dynamics-based prediction of liquefaction induced lateral spreading. *Computers and Geotechnics, 22*(3–4), 243–282.
80. Wang, Q., Liu, J., Liu, J., & Li, J., (2022). Experimental research on the impact of annular airflow on the spraying flow field: A source control technology of paint mist. *Building and Environment, 207*, 108444.
81. White, R. E., (1992). Electrochemical engineering principles by Geoffrey prentice. *Chemical Engineering Education, 26*(2), 102–103.
82. Williams, R. H., & Larson, E. D., (2003). A comparison of direct and indirect liquefaction technologies for making fluid fuels from coal. *Energy for Sustainable Development, 7*(4), 103–129.
83. Yang, Z., Yang, L., Wei, C., Wu, W., Zhao, X., & Lu, T., (2018). Enhanced nitrogen removal using solid carbon source in constructed wetland with limited aeration. *Bioresource Technology, 248*, 98–103.
84. Yeetsorn, R., Maiket, Y., & Kaewmanee, W., (2020). The observation of supercapacitor effects on PEMFC–supercapacitor hybridization performance through voltage degradation and electrochemical processes. *RSC Advances, 10*(22), 13100–13111.
85. Zhang, H. L., Baeyens, J., Degrève, J., Cáceres, G., Segal, R., & Pitié, F., (2014). Latent heat storage with tubular-encapsulated phase change materials (PCMs). *Energy, 76*, 66–72.
86. Zhang, J., Xu, S., & Li, W., (2012). High shear mixers: A review of typical applications and studies on power draw, flow pattern, energy dissipation and transfer properties. *Chemical Engineering and Processing: Process Intensification, 57*, 25–41.

CHAPTER

7 Issues for Mercury Cell Technology in Chlor-Alkali Industry

Contents

7.1. INTRODUCTION

The present chapter scrutinizes a matter that holds significant influence over the trajectory of the chlor-alkali industry in Western Europe. This analysis scrutinizes the multifaceted dimensions of a discourse that is anticipated to have far-reaching implications for North America and Asia and the global advancement of the chlor-alkali sector. The investigation encompasses environmental, economic, financial, and political considerations (Mihaiescu et al., 2012).

The primary inquiry pertains to whether or not the discontinuation of mercury cell rooms utilized in the manufacture of chlorine and caustic soda should be implemented in the immediate future. Alternatively, should the industry be granted permission to adhere to its current voluntary pledge to transition away from this outdated technology as it concludes its financial viability within two to three decades?

For over a century, the chlor-alkali industry has used the mercury cell as a primary technology for manufacturing chlorine and caustic soda. Due to historical factors, the process above became the predominant method in the European industry, whereas the asbestos diaphragm cell gained preeminence in the United States. In recent years, advancements in membrane cells have propelled them to the forefront of technological innovation. As a result, membrane cells of various types have become the preferred choice for many industries across the globe (Crook & Mousavi., 2016; Lakshmanan & Murugesan, 2914; O'Brien et al., 2005).

> **Did You Know?**
>
> Mercury cell technology was first developed in the late 19th century and became widely used in the chlor-alkali industry during the 20th century.

Historically, the use of early mercury cells resulted in significant levels of mercury emissions into the environment. Such levels would be deemed unacceptable by contemporary environmental standards. The mercury issue gained significant attention after 1959, following the association of organic mercury with the Minamata Bay catastrophe in Japan. The heightened level of public and regulatory concern necessitated prompt action. Implementing membrane technology in Japan resulted in gradually discontinuing mercury cell technology, completed by 1986, except for KOH production. This transition was facilitated by the emergence of membrane technology as a viable alternative. In other contexts, the primary objective was to intensify endeavors toward mitigating emissions, discharges, and losses of mercury from pre-existing cell rooms while expediting the advancement of membrane technology for novel cell rooms and, conceivably, for transformations. During the 1970s, European regulators raised apprehensions regarding the magnitude of mercury emissions from the chlor-alkali sector. Consequently, companies intensified their existing initiatives aimed at gradually reducing emissions (Harris, 2008; Margallo et al., 2016; Busto et al., 2011).

Despite the significant attention given to toxicity concerns in the United States during the early 1970s, the chlor-alkali industry faced comparatively lower levels of regulatory scrutiny in the region. In contrast to Europe, where there is mounting scrutiny on both asbestos diaphragm and mercury cells, the United States industry has not been subjected to the same regulatory oversight. Mercury cell rooms will probably be gradually eliminated in the United States as they end their economic viability. However, there is no indication that any regulatory measures will be implemented to hasten this transition. Similarly, asbestos diaphragm cell rooms, which constitute most of the country's capacity (as depicted in Figure 7.1), are not expected to face any regulatory pressure for replacement.

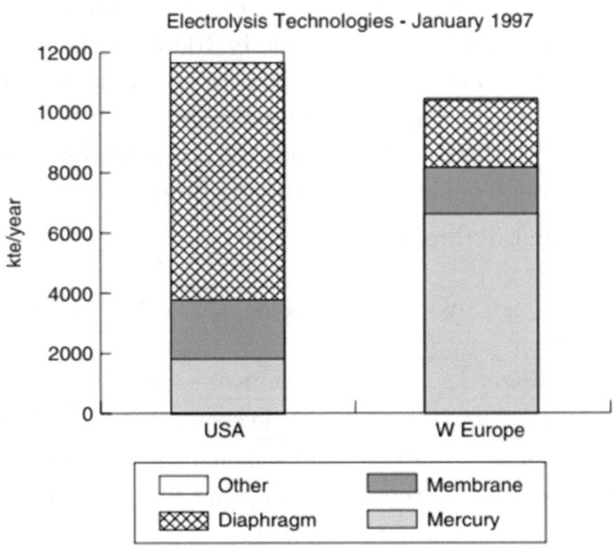

Figure 7.1. *Chlorine capacities by technology, USA and Western Europe (1997).*

Source: SRI Consulting, Creative Commons License.

The subsequent section of this chapter centers on the European context of mercury cells. This issue is of significant importance as it poses a major threat to the survival of the European chlor-alkali industry. Furthermore, there are valuable insights to be gained from the evolution of this issue, which are crucial to comprehend the type of chemical industry that will be permitted to function in the 21st century. The matter at hand is not limited to Europe alone, as the United States of America, having been at the forefront of environmental regulation between the 1970s and the 1980s, has undergone a role reversal in recent times. In the contemporary global economy and amidst the prevalence of international environmental agreements with worldwide reaches, developments in Europe – whether progressive or overly cautious – will likely be replicated in America and Asia shortly.

7.2. ENVIRONMENT

7.2.1. Occurrence and Sources

Mercury is an element that occurs naturally. The quantities of mercury released into the environment through natural means, such as from ore deposits and volcanic activity, have been estimated to range from 2,500 to 5,500 tons/year. These figures are comparable in scale to the quantities of mercury released through human activities, projected to range between 3,600 and 4,100 tons/year globally. Approximately 30,000 metric tons of mercury are present in the environment, specifically in the atmosphere or the mixing region of the oceans (Lakshmanan & Murugesan, 2014; Walker, 2016). The uppermost parts of the land mass contain tens of millions of metric tons of mercury, while the deep oceans contain even more, as indicated in Table 7.1.

Table 7.1. Natural and Anthropogenic Sources of Mercury

Typical Quantity Emitted into the Atmosphere (tons per annum)			
Natural Sources		**Anthropogenic Sources**	
Windblown dust	50	Coal combustion	2100
Sea salt spray	20	Lead production	10
Volcanoes	1000	Copper/nickel production	120
Forest fires	20	Refuse incineration	1200
Continental particulates	20	Fuel wood combustion	180
Continental volatiles	610	Chlor-alkali	7
Marine sources	770		

Total	**2490**	**Total**	**3617**

Source: M. Harris, Creative Commons License.

Measures must be taken to tackle anthropogenic emissions. To accomplish this task, these entities' origins must be known. Undoubtedly, from a historical perspective, the chlor-alkali industry has been recognized as one of the most significant anthropogenic sources. Nevertheless, the sector has significantly decreased its emissions by a factor of 10 and is no longer deemed a significant contributor. The chlor-alkali industry in Western Europe presently emits approximately 9 tons of pollutants per year, which is decreasing. This accounts for roughly 0.2% of all human-caused emissions (Zhang et al., 2021; Busto et al., 2011; Reis et al., 2009).

7.2.2. Emissions

What is the extent of mercury emissions in the chlor-alkali industry today? These are the fundamental inquiries that need to be addressed. What measures can be taken to decrease them? What are the potential hazards posed by these entities to human health and the environment?

Previously, the sequence of inquiries would have been altered, prioritizing evaluating the possible adverse effects on human health and the environment over implementing measures to reduce emissions. Refraining from causing any negative impact during that idyllic period was considered satisfactory. In contemporary Europe, a societal framework unequivocally established the objective of perpetual emissions reduction with minimal regard for the potential outcomes functions. Economic prosperity has enabled us to establish environmental objectives founded on political discourse, such as "zero risk," "zero emissions," and the complete elimination of hazardous substances. Could someone identify a substance that poses a risk to mothers? In its liquid state, inhaling pure water can significantly threaten life. Certain substances, such as antineoplastic drugs utilized in cancer treatment, are considered to be highly toxic and pose a significant risk to human health and safety. Pursuing these objectives is occurring irrespective of any potential harm or depletion of resources that may impede the resolution of pressing issues faced by contemporary European society, such as social discomfort, criminal activity, personal and communal wellness, and education (Mihaiescu et al., 2012; Crook & Mousavi, 2016; Harris, 2008).

Regarding the developing world, it is important to take this into account. Pursuing the 'Precautionary Principle' holy grail is being carried out without adequate prioritization or consideration of proportionality. Since mercury is a pervasive environmental contaminant, Resources must be allocated toward mitigating the substantial emissions from developing nations and economies in transition. The emissions disclosed by a solitary chlor-alkali plant in Siberia last year have surpassed the residual emissions emanating from every European plant until their closure or conversion (Grönlund et al., 2005; Garcia-Herrero et al., 2017; Landis et al., 2004).

Did You Know?

Mercury cell plants were often referred to as "mercury houses" due to the large amounts of mercury that contained.

7.2.2.1. Emissions From the European Chlor-Alkali Industry

Over the last 20 years, the chlor-alkali industry in Europe has significantly decreased its emissions. The industrial sector discharged 221 metric tons of mercury in 1977, proportional to 27 grams/ton of mercury cell chlor-alkali capacity (g/te). In 1998, the quantity above was diminished to slightly more than 9 metric tons, commensurate to approximately 1.5 grams per metric ton (*refer* Figure 7.2).

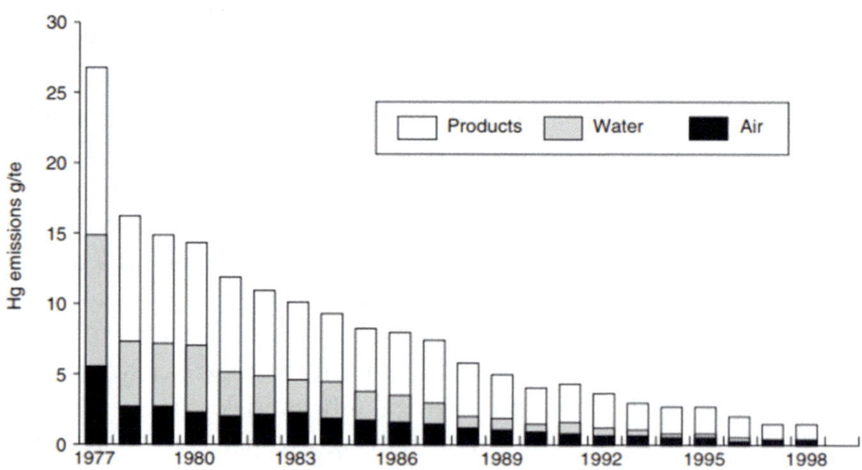

Figure 7.2. *Mercury emissions from the Western European chlor-alkali industry 1977–1998.*

Source: M. Harris, Creative Commons License.

The figures above denote a significant decrease of 98% in product losses, 99% in water discharges, and 91% in atmospheric emissions. It is unlikely that any other significant industry can rival the extent of emissions reductions achieved by the measures in this context. It is, therefore, reasonable to question the rationale behind the considerable government resources allocated to expediting the discontinuation of mercury cell rooms. It is worth noting that the industry acknowledges the impracticality of installing new capacity based on this outdated technology. The thought process of the regulator can often appear to be a complex and perplexing phenomenon, akin to a puzzle shrouded in obscurity and ambiguity (Garcia-Herrero et al., 2017; Margallo et al., 2016).

7.2.2.2. Wastes

A disputed matter in the northern European region pertains to solid wastes containing mercury, including sludge generated during routine operations and waste produced during the closure and demolition of outdated cell rooms (Scheme, 2010; Grönlund et al., 2005). In numerous nations, regulatory bodies appear content with the notion that these waste materials do not constitute transfers to the environment unless considered from a geological perspective. The industry practice documentation is comprehensive and subject to the scrutiny and authorization of regulatory authorities at both national and local levels. In most West European nations, retrieving all retrievable mercury from waste materials is accomplished through retorting. At the same time, any remaining residues are appropriately disposed of in registered landfills or other underground deposits (Garcia-Herrero et al., 2017; Wängberg et al., 2003). This disposal process typically involves some form of inerting treatment, such as conversion to sulfide or encapsulation in mineral materials, following regulations commensurate with the specific levels of mercury present in the waste materials. Various monitoring programs are implemented to reinforce regulated disposal practices, such as assessing mercury levels in the air or groundwater near the sites. Certain nations consider that disposing of mercury in secure underground storage constitutes an environmental emission, even though the methods employed involve restoring the mercury to its original location. The substance originated as a form of mercury ore that underwent extraction and refining processes, likely conducted in other regions of Europe (Lee et al., 2018; Crook & Mousavi, 2016; Brinkmann et al., 2014).

7.2.2.3. End-of-Life Issues

An additional plausible origin of mercury emissions exists, which surpasses all discourse on emissions, discharges, and losses. However, until recently, it was not under regulatory scrutiny. The operational cell rooms in Western Europe are estimated to contain approximately 12,000 metric tons of pure mercury. The fate of the mercury upon closure of the cell rooms is uncertain. The data above signifies approximately 1500 years of emissions from functional cell rooms and roughly 15 years of global mercury output at current rates. The resolution of this matter is of significant importance for environmental concerns and the mercury mining sector (Southworth et al., 2004; Lee et al., 2018; Kinsey et al., 2004).

Certain Northern European authorities have suggested depositing mercury in underground mines with appropriate geological characteristics. In a somewhat erratic manner, they also decline to engage in the import or export of mercury, thereby hindering the realization of this objective. Upon reflection, it becomes evident that implementing deep-mine storage necessitates extracting and processing fresh mercury ore to meet the valid worldwide demand for mercury. Notably, a significant proportion of global mercury production, exceeding 50%, is extracted from within the confines of the European Union, which presents an ironic circumstance. The prospect of Sweden engaging in the burial of pure mercury while Spain is actively involved in the mining and smelting of fresh mercury from ore could potentially serve as a source of inspiration for a forthcoming environmental caricaturist. What is the current status of recycling and resource conservation efforts?

Euro Chlor opines that a more logical and productive strategy is being formulated through voluntary collaboration with the mercury mining sector. Drafted proposals have been formulated on the premise that the mercury found in cell rooms would be restored to the mercury mine, thereby serving as a substitute for the mining and smelting of fresh mercury on a one-to-one basis. The proposed initiative would effectively utilize the current infrastructure for transporting and handling mercury while preventing the introduction of additional metallic mercury into the environmental stock. The sole remaining challenge pertains to the efficiency of the logistical procedures, with a projected capacity of no more than 400–500 tons per annum. Attempts are being undertaken to augment this. However, it appears that in the event of the entire 12 000 tons being discharged within a decade – as would be the case with a phase-out date of 2010 – it would regrettably be necessary to recommence mercury mining and smelting activities. Alternatively, suppose cell rooms are decommissioned upon reaching the end of their economic viability. In that case, it is plausible that mining activities in Europe may experience a hiatus of approximately two decades, followed by a resumption at significantly reduced levels that would suffice to satisfy the demands of a substantially diminished global market. The decision ultimately lies within the purview of governmental authorities (Garcia-Herrero et al., 2018).

7.2.3. Environmental Impact Conclusion

According to recent studies, which include comprehensive risk assessments conducted by Euro Chlor, it has been determined that contemporary chlor-alkali processes across Europe do not pose any discernible risk to either individuals or the ecosystem. Subsequent measures would be predicated solely on a precautionary methodology. It is widely acknowledged within the industry that additional measures are necessary, given the prevailing social and political climate in Western Europe. The organization is dedicated to achieving additional reductions in emissions and refraining from implementing any supplementary capacity that relies on mercury cell technological advances. Numerous scenarios involve the amalgamation of reducing emissions and the closure of extant plants. To evaluate the environmental advantages and disadvantages of various scenarios scientifically, it is essential to acknowledge that the significant quantity of mercury in the ecological cycle and its sluggish transfer rates to geological reservoirs, such as deep-sea sediment, must be considered. Therefore, the environmental assessment of a phase-out scenario cannot be based solely on the emission rate in a particular year or the duration of emissions until the phase-out is accomplished. The cumulative environmental burden, which refers to the quantity of mercury introduced into the environment before the phase-out period, is a product of the convolutions of these two parameters. The significance of this notion will be elucidated as the succeeding parts of this chapter delve into the various options for phase-out strategies (Lee et al., 2018; Harris, 2008).

The transformation and closure of approximately 60% of the chlor-alkali industry in Western Europe is a significant endeavor, regardless of the approach taken. Subsequently, both social and economic aspects are discussed (Campanari et al., 2019; Jung et al., 2014; Guandalini et al., 2017).

Remember:

The discovery of Minamata disease outbreak in Japan in the 1950s, caused by mercury pollution from a chlor-alkali plant, played a significant role in raising awareness about the health and environmental risks of mercury cell.

7.3. ECONOMICS

Environmentalists may be inclined to overlook the domains of economics and finance. According to popular belief, the environment holds greater significance than monetary value. Nonetheless, there is a desire to enhance and safeguard the environment in practical settings where the general populace resides. However, such endeavors entail financial expenses, and it is acknowledged that finite resources necessitate prioritization in terms of expenditure. It is acknowledged that generating profits is imperative for the sustenance of businesses. Therefore, no justification is needed for discussing the economic aspects of cell room conversion (Guo & Sun, 2020; Moussallem et al., 2008; Anderson et al., 1990).

Similar to other significant capital investment endeavors, the financial justification for this project is predicated on the financial gains, as measured by various metrics, and the accessibility of funding (Zhang et al., 2022).

7.3.1. Capital Cost of Conversion

Several approximations exist of the capital expenditure associated with the conversion of cell rooms, which unavoidably exhibit discrepancies across different facilities, even when normalized by capacity. Significant discrepancies in reported estimates can result from the traditional 'boundary problem,' which pertains to the inclusion or exclusion of certain factors. Numerous examples exist, and the primary objective of this chapter is not to delve into the technical and economic intricacies of converting cell rooms. The expenses involved in transforming a standard mercury cell room comprise not solely the expenses related to the electrolyzers but also, at a minimum base level, the interconnected capital expenditures linked to caustic recycling, caustic evaporation, secondary brine purification, dechlorination, and supplementary gas cooling. In numerous instances, supplementary capital may be necessary for installing novel rectifiers (and conceivably new electrical supply and switchgear), modifications to the provision of utilities and gaseous HCl, novel edifices, decommissioning, and site remediation. The literature presents a range of capital costs for chlorine capacity, ranging from less than \$400 to greater than \$800 per ton. The costs presented are denominated in US dollars, primarily due to their approximate equivalence to euros at the conversion rates of the early 2000s. Suppliers highly motivated to provide membrane cell equipment may offer a quotation that falls within or below the lower end of the price range. However, it is important to note that this pertains solely to the expense of the electrolysis apparatus (Wang et al., 2014; Leddy, 1980; Hurskainen & Ihonen, 2020). The computation is additionally intricate, given the trade-offs between capital and operating expenditures. Optimizing for high current density during the design phase may reduce the initial capital expenditure. However, this approach may lead to an increase in both energy consumption and operating expenses. The emergence of lower voltage membranes may influence the trend toward higher current densities. However, it remains to be seen whether regulatory measures to reduce energy costs and consumption will counteract this trend and result in higher capital costs (Anderson et al., 1990; Hofmann et al., 2022).

In a customary conversion scenario, the necessary capital is expected to fall within $400 to $700 per ton of chlorine capacity. The extant chlor-alkali industry in Western Europe necessitates a capital outlay of approximately $3.2–4.8 billion. To attain phase-out by 2010, the capital above must be expended within the next decade, specifically at an annual rate ranging from $320 million to $480 million over the next 10 years. The amount above is approximately equal to the collective capital budget of the West European industry. In other words, it would preclude any efforts towards restructuring and modernization, expansion of production capacity to meet the demands of downstream markets, substantial maintenance of current facilities, and allocation of funds towards safety, health, or environmental initiatives.

The inquiry from the industry pertains to the alignment of the desires of even the most fervent environmentalist with the current scenario wherein the industry has made an unambiguous commitment to phase out within a reasonable timeframe. Additionally, the industry has successfully reduced mercury emissions to an environmentally safe level and has crucial environmental enhancement initiatives it intends to undertake. Therefore, the question arises whether this utilization of resources is socially responsible and sustainable (Klaucke et al., 2023).

7.3.2. Energy Savings

Advocates of swift phase-out have frequently asserted that the substantial expenses associated with conversion are recuperated expeditiously through diminished operational expenses, particularly using decreased energy usage. This assertion appears to be more straightforward than providing evidence. The adoption of membrane technology has resulted in significant reductions in electrical energy consumption, with these savings increasing over time as the technology has progressed. The purported savings are not as substantial as previously asserted and are significantly counteracted by augmented energy demands in other stages of the process, particularly in cases where thermal energy is necessary for the concentration of caustic liquor. As is customary, the precise equilibrium will be heavily dependent on the particularities of the site. Typically, the primary factor determining the existence and magnitude of energy conservation is the origin of the electrical energy supplied to the location (Li et al., 2014; Wang et al., 2014; Guo et al., 2020).

Multiple literature surveys indicate that a standard conversion has the potential to achieve energy savings ranging from 7% to 14% within site boundaries when compared on a like-for-like basis. This information can be found in Table 7.2.

In a comprehensive analysis of the energy economics of current electrolysis plants, Dr. I. Boustead, a renowned expert in eco-profiles, has adopted a more expansive perspective on the energy comparison between mercury and membrane technologies. The present study indicates that the energy benefit of membrane cell technology is limited to a mere 7% compared to mercury cells. However, this estimate may be inflated because membrane cells are primarily located in regions that employ more effective methods of electricity production (Park et al., 2019; Moussallem et al., 2008; Khasawneh et al., 2019).

The cost savings range related to the conversion process appears to be within the range of \$30 to \$50 per metric ton of chlorine.

7.3.3. Financial Case

A comprehensive financial analysis must consider capital expenditures, potential energy savings, operational cost reductions, and financial and tax implications (Garcia-Herrero et al., 2017; Chen & Murata, 2017). The specific details will vary depending on the site.

Table 7.2. A Comparison of Energy Consumption: Mercury versus Membrane

	Mercury Technology	**Membrane Technology**
Theoretical voltage (V)	3.15	2.19
Cell voltage (V)	3.9–4.2	3.0–3.6
Current density ($kA\ m^{-2}$)	8–13	3–5
Caustic strength (wt.%)	50	33

AC electrical power consumption (kWh ton^{-1} Cl_2)	3360 at 10 kA m^{-2}	2750 at 3.6 kA m^{-2}
Other electrical power consumption (kWh ton^{-1} Cl_2)	200	140
Steam consumption for caustic evaporation (kWh ton^{-1} Cl_2)	0	180
Total energy consumption (kWh ton^{-1} Cl_2)	3560	3070

Source: John Moorhouse, Creative Commons License.

The fundamental financial variables associated with transforming mercury to membrane exhibit a relatively distinct range in broad scope and encompassing diverse circumstances. To account for the broad spectrum of published estimates, an analysis of the financial feasibility could be conducted using capital costs ranging from $400 to $700 per ton of chlorine, functioning cost savings ranging from $30 to $50 per year per ton of chlorine, and discount rates ranging from 8% to 14%. Table 7.3 presents a range of net present values (NPVs) obtained through conventional financial modeling (Jia et al., 2016).

Table 7.3. Financial Parameters: Mercury-to-Membrane Conversion

The Capital Cost of Conversion ($m per ton of Chlorine Capacity)	**Operating Cost Savings from Conversion ($ per year per ton of Chlorine Capacity)**	**Discount Rate (%)**	**NPV ($m)**	
			Best Case	**Worst Case**
Range: 40, 55, 70	40	12	–10	–37
55	Range 30, 40, 50	12	–16	–29
55	40	8–14	–11	–26
40 (best) or 70 (worst)	30 (worst) or 50 (best)	8 (best) or 14 (worst)	+6	–38

Source: John Moorhouse, Creative Commons License.

The findings indicate that the project's value is predominantly negative across all instances. The financial feasibility of the case can only be achieved when discount rates are less than 10%. This is possible only when the lowest possible capital cost of $400 million and the highest possible savings rate of $50 per year are assumed simultaneously. The prevailing discount rates for significant capital projects within the chemical industry are presently at or above 12%, with rates exceeding 15% reserved for ventures with higher risk. It should be noted that the project is not expected to yield any discernible environmental benefits.

What is the rationale behind performing conversions? The solutions are straightforward. Upon concluding its economic operating lifespan, typically 40–50 years, a mercury cell room must either undergo replacement or closure. The implementation of cutting-edge substitutes necessitates a transition to membrane-based technology. There is a certain degree of variability regarding the economic lifespan of a plant, which is contingent upon the specific circumstances of each case. The exact timing of conversion or closure is contingent upon a company's comprehensive strategy, its array of manufacturing operations, its downstream users of chlorine and caustic, its degree of integration and ability to attain in-house economics, and its vision for its standing in an industry that is currently undergoing significant restructuring and rationalization, particularly in Western Europe. The project's feasibility will also be contingent upon the accessibility of capital, which may be subject to competition with other capital demands within the overarching corporate survival strategy (Kumar & Singh, 2022).

In summary, converting cell rooms occurs either due to the deterioration and subsequent replacement of the existing facilities or as a component of a corporation's comprehensive reorganization and growth strategy, wherein a feasible financial rationale may exist, and adequate funding may be secured. Conversions or closures are expected over the next few decades as European industry deforms and enhances its global competitiveness. The efficacy of governmental pressure in expediting the pace of cell room conversions remains uncertain. Nevertheless, it is plausible that the government's actions may compel closures instead of conversions, resulting in various social and economic ramifications (Otashu & Baldea, 2019).

7.3.4. Macroeconomic Effects

After introducing the topic of compelled cessation, it is pertinent to examine the broader economic ramifications briefly. Forecasting the macroeconomic impacts of the closure of a significant portion of the chlor-alkali industry in

Western Europe is challenging. The foremost ramifications of this closure are expected to manifest in the form of downstream industries' ripple effects. These industries will have to operate in extremely competitive markets for brought in caustic and EDC, which will be in short supply. It is probable that a significant number of key consumers within the chlor-alkali sector, specifically those involved in the PVC chain, would progressively relocate from Western Europe to be closer to their supply sources (Zhou et al., 2009; Hosseini & Zardari, 2017).

Consequently, a persistent impact is anticipated for several decades, whereby commodities situated increasingly farther downstream from chlorine would necessitate importation into Western Europe rather than being domestically produced. The potential macroeconomic and strategic implications of Western Europe's gradual shift away from its manufacturing sector and towards a greater reliance on service industries is a topic that could be explored in a thought-provoking, albeit sad, literary work. However, it is not the focus of this particular chapter.

In the immediate future, there will be employment ramifications that are more immediate and direct. Due to its capital-intensive nature, the chlor-alkali industry does not exhibit significant employment levels. Nevertheless, the closure of a business has significant indirect employment ramifications. Governmental estimates could be considered more reliable than industry estimates due to their perceived higher level of information. According to the European Commission's estimation, the chemical industry in the European Union is expected to experience a job loss of 10,000 by the year 2010 without considering any secondary effects. According to recent communication from the German government to OSPAR, the chlor-alkali industry in Germany, which is presently reliant on mercury technology to 50%, employs 6,650 individuals. Notably, this industry provides livelihoods to approximately 2,50,000 individuals in Germany alone (Goodman, 1994; Bassi & Yudken, 2009).

The task of measuring the macroeconomic impacts of cell room closure is a highly challenging one. The emergence of this challenge is noteworthy because the chlor-alkali sector serves as the foundation for a considerable portion of various manufacturing industries, consequently contributing to the tax revenue of national governments (Whitfield & Brown, 2014).

7.3.5. The OECD Analysis

By implementing environmental accounting, numerous individuals contend for incorporating environmental factors into commercial and financial strategies. This approach endeavors to quantify economic activities' favorable and unfavorable environmental impacts. Undoubtedly, achieving this objective is highly valuable, albeit challenging. The risk-benefit analysis is a crucial component in the process of making calculations. It is a cost-benefit analysis that evaluates the balance between the advantages and disadvantages of a particular undertaking. Given the current levels of mercury emissions, it is improbable that any health or ecological effects would arise. However, a retrospective examination of past events is necessary for a comprehensive worst-case analysis. An instance of a case study was presented to the OECD by Japanese scholars in January 1998.

The study was conducted after replacing all significant mercury cell rooms with membrane technology following the Minamata incident. The present investigation computes the benefit-to-risk ratio (BRR) for diverse scenarios by employing loss of life expectancy (LLE) as a metric to compare different types of health hazards. BRR is defined as the ratio of costs to risk reduction.

The research findings suggest that the government ruling to ban the utilization of the mercury electrode process lacked cost-effectiveness. The study indicates that had the industry replaced the mercury process during the renewal of existing plants, the BRR value would have significantly decreased. To avoid formulating an excessively stringent policy, the government must relinquish its stance of absolute risk aversion. The research conducted a cost analysis to determine the estimated expense of prolonging an individual's life expectancy by one year, which resulted in a figure of $57 million. This value was then compared to other environmental health policies implemented in Japan, which ranged between $0.24 million and $0.45 million per life-year, and toxic control policies enforced by the US Environmental Protection Agency (EPA), estimated to cost $4.2 million per life-year (Campanari et al., 2019; Bosch et al., 2009).

The outcomes are self-evident. There exists no justification, either from an environmental or economic perspective, to expedite the process of converting mercury cell rooms beyond the previously agreed-upon time frame of 20–25 years, as has been pledged by the European industry. The sole justification appears to be political, leading us to the concluding segment of this examination (Le Faucheur et al., 2016; Garcia-Herrero et al., 2017; Li et al., 2014; Guandalini et al., 2017).

7.4. REGULATION

Several decades ago, an environmental argument existed for implementing more stringent regulations and monitoring mercury emissions, discharges, and losses from the chlor-alkali industry, including in Western Europe. The industry had conformed to the regulations that were in place during that period and had modified its operational procedures to mitigate substantial hazards to human health. However, a comprehensive reaction to the emerging apprehensions regarding potential environmental impacts, specifically those that could occur at remote locations from the facilities, had not yet been provided. The emergence of worldwide pollutants instigated apprehensions regarding the stratospheric ozone layer, ultimately resulting in the implementation of the Montreal Protocol. Despite their potential lack of efficacy and misguided nature, most contemporary regulatory endeavors can be traced back to the apprehensions of previous eras (Brinkmann et al., 2014; Bissett & McBeath, 1988).

The industry can learn a noteworthy lesson from this observation: mere compliance with regulations is inadequate; instead, the industry must proactively evaluate the potential for harm stemming from its operations and undertake voluntary measures without relying on governmental response. If no action has been taken, governments will probably react excessively. The approach in question has been denoted as Responsible Care, while environmentalists have referred to it as the Precautionary Principle. A potential compromise in terminology could be Sustainable Development. However, it is noteworthy that the chlor-alkali industry in Western Europe is currently subject to regulation in at least three significant international arenas (Scheme, 2010; O'Brien et al., 2005; Ashworth et al., 1987).

7.4.1. UN/ECE

The United Nations Economic Commission for Europe (UN/ECE) formulated the Convention on Long-Range Transboundary Air Pollution in 1979 to address the mounting apprehensions regarding air pollution that could traverse national boundaries and cause unfavorable impacts in regions far from its origin. The UN/ECE's geographical coverage extends beyond Europe, including all territories of the former Soviet Union, the United States, and Canada, despite its name being historically rooted in Europe. The protocols associated with this Convention pertain to the contamination caused by various substances such as nitrogen oxides, sulfur dioxide, volatile organic compounds (VOCs), persistent organic pollutants (POPs), and heavy metals. Currently, the heavy metals covered by the protocols are cadmium, lead, and mercury. The Heavy Metals Protocol acknowledges the chlor-alkali sector as a noteworthy contributor to releasing mercury into the environment. In certain regions undergoing economic transition, this assertion remains true (Hackl et al., 2001; Rallo et al., 2012; Pacyna et al., 2006).

The Protocol mandates:

- Adoption of efficient measures to reduce emissions from the levels recorded in 1990.
- Implementing the best available technology (BAT) is expected to occur within two years for new plants and eight years for existing plants following the Protocol's entry into force. This timeline is estimated to be around 2004 and 2010, respectively.
- The implementation of emission limit values (ELVs)

For mercury cell rooms it expects that the application of BAT will result in emission levels well below 2.0 g/te (of chlorine capacity). It is worth noting that certain facilities have successfully attained emission levels that are considerably below 1.0 g/te. The Protocol has established a numerical emission limit of 0.01 g/te for new plants, equivalent to prohibiting mercury technology. However, determining such limit values for existing plants will be deferred until two years after the Protocol's implementation.

All chlor-alkali plants in Western Europe currently comply with the regulations outlined in this Protocol. Furthermore, Western European producers have voluntarily pledged to abstain from using mercury technology for new plants. The Protocol's primary significance lies in its establishment of rigorous benchmarks for manufacturers operating in Eastern Europe, specifically within the territories of the erstwhile Soviet Union. In certain instances, the emission levels in these regions have exceeded the Protocol's stipulations by over 50-fold, underscoring the Protocol's critical role in promoting environmental sustainability. Establishing a level playing field between North America and Western Europe is also valuable (Scoullos et al., 2001; Liu et al., 2010; Kolon et al., 2015).

7.4.2. IPPC Directive and the Chlor-Alkali BREF

The 1996 IPPC Directive serves as the primary instrument for regulating the control and prevention of pollution within the European Union. The Member States mandated that the Directive be transposed into national laws and regulations by September 1999. The Directive is firmly grounded in the concept of BAT, which is meticulously defined within the text. While the Directive does not establish specific emission limit values, it provides protocols that define the measurable emissions levels likely to arise from implementing BAT. The nuanced nature of this strategy facilitated the consensus of all Member States regarding the Directive. However, it has subsequently engendered significant discourse concerning the distribution of authority between the competent regional bodies (which possess the capacity to establish emission thresholds on a plant-specific level) and the European Commission, which serves as the custodian of BAT as delineated in the Directive (Brinkmann et al., 2014; Aldaco et al., 2010).

The directive is implemented in practice through the creation of BAT reference documents (BREFs), which function as instructional materials for specific industrial sectors (e.g., the chlor-alkali industry) or for crosscutting activities, known as "lateral BREFs."

The inception of the BREF for the chlor-alkali industry commenced in December 1997. It persisted until late 1999, characterized by a sluggish pace and a partial comprehension of the intricate technicalities involved in chlor-alkali manufacturing. Hence, it was not unexpected to discover that the suggestions indicated that BAT was categorized as a membrane technology not solely for newly established facilities, which is a justifiable stance, but also for pre-existing ones. By approximately 2008, plant closure would have been equivalent to the situation in several countries where 'BAT' was being misconstrued as a uniform obligation instead of being viewed as guidance for local regulators to decide on permission for each plant (Emmott & Slayne, 2005; Lindström et al., 2003).

A significantly enhanced and expedited technical discourse occurred during the latter part of 1999 and the beginning of 2000. The Technical Working Group convened in Seville in March 2000 and widely approved the latest iteration of the BREF, which is now in its third draft. Additionally, comprehensive research is necessary. However, there is a positive outlook that the BREF will be finalized in the summer of 2000 as a mutually agreed upon document between the industry and regulatory bodies and subsequently presented at the Information Exchange Forum's (IEF) September 2000 assembly for official approval at the governmental level. The BREF document is expected to serve as a directive and significantly impact the national regulatory bodies' decisions. The industrial sector should anticipate a minimum pressure towards prematurely converting the staying mercury cell rooms. Additionally, in numerous instances, there will be the imposition of highly demanding emission limits, although some consideration will be given to the unique circumstances of each plant. The initial enthusiasm for achieving goals as low as 0.2–0.3 g/te, which is the optimal performance attained by plants constructed and designed appropriately from the outset and functioning under the most advantageous local conditions, has been tempered. However, there is an inescapable push towards attaining levels significantly below 1.0 g/te (Breedveld, 2000; Abreu et al., 2008; Masnicki, 2018).

7.4.3. OSPAR

OSPAR is a prominent and influential marine conservation convention that operates within European and close waters, among several others. The OSPAR Convention, officially called the Convention for the Protection of the Marine Environment of the North-East Atlantic, was made available for signature during the Ministerial Meeting of the Oslo and Paris Commissions in September 1992. This agreement's signatories comprise nations with territories in the north-east Atlantic region, including all the European Union member states, except Austria, Italy, Greece, Norway, Iceland, Switzerland, and the European Commission. The entity above, primarily overseen by the environmental ministries of the Contracting Parties, is deficient in the equilibrium present in EU legislation, wherein consensus is required among the environment, trade, industry, and other directorates. At the outset, the entity exhibited an atypical absence of transparency commonly observed among European institutions, whereby non-governmental organizations, be they environmental or business-oriented, was precluded from attending its sessions in an observer capacity. The situation was particularly regrettable because OSPAR possesses considerable legislative authority. However, the tension between its powers and those of the European Union, of which most OSPAR Contracting Parties are members, remains a source of intricate and, at times, contentious discussions. Since 1998, notable enhancements have been made to the functioning of OSPAR. The noteworthy level of engagement has garnered involvement from various governmental departments, and the organization has invited non-governmental organizations to serve as observers. Implementing these modifications has yielded dual advantages, namely, enhancing the legitimacy of OSPAR as an entity and augmenting the caliber of its decision-making process by integrating empirical technical data before the formulation of regulations (Raldúa et al., 2007; Reis et al., 2009; Ullrich et al., 2007).

The elements and compounds of Mercury exhibit toxicity and persistence and possess a notable capacity for bioaccumulation. As per the statement, despite the natural existence of mercury, its contamination has consistently remained a significant concern for OSPAR. The chlor-alkali sector has garnered significant scrutiny, partly due to its historical environmental record and partly due to the chemical industry's vulnerability as a target compared to other major mercury sources such as combustion processes (e.g., power generation and incineration).

Initially, regulatory measures were primarily directed toward managing effluent discharges into the aquatic ecosystem. The choices above and suggestions have been rendered obsolete due to the implementation of more rigorous European Union regulations and the proactive measures taken by industries to mitigate mercury emissions. Consequently, the direct discharge of mercury into the hydrosphere has been significantly reduced, diminishing its impact as a concern. By 1990, the focus had shifted toward the atmospheric discharges from the chlor-alkali sector, which were deemed unacceptably elevated during that period. The Paris Commission, one of OSPAR's predecessors, made Decision 90/3 in a private meeting in 1990. The decision included the following:

- A requirement for mercury-based chlor-alkali plants to adhere to an emissions standard of 2 g/te for atmospheric emissions by December 31, 1996.
- Additionally, the contracting parties suggested that current mercury cell chlor-alkali plants should be phased out as soon as possible. The goal is to eliminate them by the year 2010 fully.

The initial prerequisite has been fulfilled and surpassed. Considerable debate has arisen regarding the second requirement due to the convoluted legal language used, particularly about whether it constitutes a recommendation within a decision. Despite the uncertain legal efficacy of this provision in PARCOM 90/3, it carries significant political weight, and a majority of the signatory nations recognize the importance of adhering to the essence of the Recommendation. Regrettably, variations in the interpretation of regulations by different nations do not promote a fair and equitable environment within the European Union, a fundamental principle of the Treaty of Rome. This situation also causes significant ambiguity within the industry (Gadelha et al., 2010).

The significance of these variations in understanding and uncertainties would be diminished if not for the ecological and financial dimensions of the matter expounded in the earlier segments of this chapter. If the Recommendation were implemented strictly and all chlor-alkali plants in the OSPAR area were compelled to shut down or transform by 2010, it would result in significant economic and social harm without any discernible environmental benefit. EU legal experts may wish to contemplate the circumstances surrounding prominent producer nations such as France and Spain, wherein certain facilities fall under the jurisdiction of the OSPAR catchment area. In contrast, others discharge into the Mediterranean basin. Because industrial interests were entirely omitted from the discussions that culminated in the 1990 resolution, it is possible that the industry was somewhat sluggish in acknowledging the gravity of the threat posed by PARCOM 90/3. As previously outlined, significant progress was achieved in reducing emissions by implementing abatement measures (Yarime, 2007; Mukherjee et al., 2004).

Additionally, there were ongoing efforts to convert and close certain operations. However, it became evident in the latter half of the decade that achieving the 2010 target was unfeasible regardless of the cost and would not yield any further ecological advancements. Certain governmental bodies also experienced growing unease regarding the 1990 resolution. Thankfully, the OSPAR procedure includes a compulsory decennial evaluation of determinations and suggestions, and the current year is 2000.

The initial course of action OSPAR took was a conventional governmental measure of deferring the evaluation of PARCOM 90/3 until 2001. In contrast to the private sector, governmental entities possess the authority to establish regulations and may postpone their adherence (Rallo et al., 2012; Gibičar et al., 2009; Erdmann & Sartorius, 2005). However, the matter has garnered significant focus within OSPAR from most of its member states (Kirk & Silfverberg, 2006). The discourse is imbued with a distinct flavor, considering the inclusion of mercury and its derivatives in OSPAR's roster of 15 "Priority Substances" in 1998, as well as the declaration of the Sintra Statement by OSPAR Ministers during the same year, which incorporates the text above.

7.4.4. Working Within OSPAR

Following acknowledging certain issues regarding the interpretation and execution of PARCOM 90/3 during the 1997 Commission meeting, OSPAR facilitated an opportunity for the industry to elucidate the measures taken to address the raised concerns and the remaining tasks and their corresponding approaches. Euro Chlor, the trade association representing the industry, formulated a strategic plan to address the issue of mercury plants during their operational phase, during and after their conversion, and eventual closure.

- Euro Chlor has identified, through rigorous scientific research conducted in partnership with various academic institutions and global specialists, that the current emission levels have significantly reduced to a point where the probability of any environmental impact is highly improbable.
- Furthermore, in the unlikely event of any impact, the magnitude of such impact would be negligible.
- The extent of the environmental impact was contingent upon the total quantity of mercury released from the present until the cessation of conversion/closures.

This served as the basis for implementing a shut-down approach for the current mercury cell chambers. Euro Chlor faced the challenge of achieving significant reductions in emissions such that the cumulative mercury burden would be lower than the optimal level attainable under the 2010 phase-out scenario, even after the exit from mercury as plants approached the end of their economic life. The task at hand proved to be challenging. However, it expeditiously became apparent that the cessation of operations in 2010 was improbable and that any endeavor to accomplish it would wholly redirect resources from additional endeavors to reduce emissions. In practical application, a compromise would need to be made between reducing emissions and closure timing. The specifics of this matter were ascertained via autonomous research endeavors conducted by SRI Consulting, which devised a comprehensive framework encompassing the minutiae of individual plants for the gradual phasing out of mercury technology. By integrating this model with established emission levels on a per-plant basis, it became feasible to compute mercury loads for various situations (Raldúa et al., 2007).

Concurrently with the establishment and execution of optimal procedures for all facets of the process, the endeavors above were carried out in tandem with the advancement and execution of best practices for the operation, decommissioning, and demolition of mercury cell rooms. Special emphasis was given to the atmospheric discharges, which currently hold a prominent position, and subsequently to various intricate matters related to garbage management and, notably, the suitable destiny of the 12,000 metric tons of uncontaminated mercury present in functional cell chambers.

OSPAR facilitated the technical endeavors by establishing an Intersessional Correspondence Group (ICG) comprising a group of proficient individuals from industry and government sectors. The ICG provides reports on point sources of pollution, commonly referred to as POINT, to OSPAR's annual meeting. This endeavor's culmination was a two-day chlor-alkali workshop, known as 'WOCAI,' held in Madrid in September 1999. During the workshop, comprehensive technical presentations were delivered on all aspects of mercury matter. The workshop's findings partially agreed with the industry's perspectives. In cases where technical discrepancies persisted, they were referred to the ICG for resolution (Ullrich et al., 2007).

7.4.5. Six Voluntary Commitments

A crucial advancement in fostering a more effective and positive rapport with the OSPAR officials has been the growing acknowledgment within the industry that a strategy founded on robust scientific principles and practical technological approaches is indispensable, albeit not exhaustive. Regardless of personal preferences, it is a widely acknowledged fact that individuals tend to lack logical reasoning skills and possess limited knowledge in scientific and technical domains. The complexity of the real world is characterized by many facets whereby diverse views and emotions influence behaviors and decisions. Elected officials and the regulatory bodies they oversee are tasked with representing and responding to the realities of the world. The task at hand can be executed with varying degrees of proficiency. Individual preferences may vary.

Due to this rationale, the Chief Executive Officers of each chlor-alkali producer in Western Europe have voluntarily committed to six binding agreements through Euro Chlor. All except for one of these are without conditions. From this perspective, it can be inferred that these actions are not a component of any negotiation process. Rather, they serve as a means of showcasing the industry's unwavering dedication to addressing public apprehensions regarding its utilization of the mercury process, regardless of its historical context (McHugh et al., 2016).

The Madrid Commitments were officially declared during the OSPAR workshop held in Madrid in September 1999. To summarize, the items above can be listed as follows:

- The production capacity of mercury chlor-alkali will not experience any augmentation. This statement represents an unambiguous reaffirmation of a pledge originally made in 1995.

The statement above denotes an implicit pledge to gradually discontinue using mercury cell rooms as they conclude their operational lifespan.

- Mercury cells shall not be dispatched to tertiary entities for reuse. This offers a sense of confidence that technological advances will not be relocated solely beyond the confines of Western Europe.
- There will be a reduction in the emissions of mercury. Despite having fulfilled all the prerequisites of PAR-COM 90/3 about this matter, the corporations pledge to exceed the established standards and strive for ongoing enhancement.
- The authorities will receive comprehensive data on mercury emissions, subject to independent third-party auditing, and will be presented on a plant-by-plant basis with complete transparency. Paradoxically, this surpasses the extent of disclosure that certain governments have been inclined to provide thus far, given that certain nations have resisted divulging the efficacy of specific facilities during global deliberations.
- The mercury cell rooms that are still operational will either undergo conversion to non-mercury processes or be closed down upon reaching the end of their financial sustainability. The precise timing of the event is contingent upon the accessibility of financial resources and macroeconomic variables that are predominantly subject to the influence of governmental entities rather than the industry. According to various independent analyses, it can be inferred that the mercury process in Western Europe is likely to come to a halt in the 2020s.
- The disposal of mercury during shutdown procedures will be executed in a manner that does not increase emissions to the surrounding environment. The management, conveyance, and elimination of waste materials will be executed following optimal environmental protocols and the most advanced methodologies currently accessible. The 12,000 metric tons of elemental mercury obtained from the cell rooms will undergo a recycling process. The primary priority is to implement re-use strategies in the remaining cell rooms to prevent any escalation in the mercury inventory of the industry. To minimize the anthropogenic inventory on a global scale, any remaining amount will be directed towards the mercury mines to.

7.5. SUMMARY

Mercury cell technology has been historically used in the chlor-alkali industry for the production of chlorine and caustic soda. However, this technology has several significant environmental and health issues associated with it. One of the most important significant concerns with mercury cell technology is the release of mercury into the environment. Exposure to mercury can have serious health effects on humans. Due to the concerns with this technology many countries have implemented strict regulations and restrictions on its use. The chlor-alkali industry has been gradually transitioning to alternative technologies such as membrane cell and diaphragm cell technologies.

REFERENCES

1. Abreu, S. N., Soares, A. M. V. M., Nogueira, A. J. A., & Morgado, F., (2008). Tree rings, *Populus nigra* L., as mercury data logger in aquatic environments: Case study of an historically contaminated environment. *Bulletin of Environmental Contamination and Toxicology, 80*, 294–299.
2. Aldaco García, R., Onandía de Dios, R., & Irabien Gulías, Á., (2010). *Impact Analysis of the Spanish Electric Energy Market Liberalization on Chlor-Alkali Industry*.
3. Anderson, E., Taylor, E. J., Vilambi, N. R. K., & Gelb, A., (1990). Preliminary analysis of fuel cell derived technologies applied to energy conservation in the chlor-alkali industry. *Separation Science and Technology, 25*(13–15), 1537–1554.
4. Anirudhan, T. S., & Shainy, F., (2015). Effective removal of mercury(II) ions from chlor-alkali industrial wastewater using 2-mercaptobenzamide modified itaconic acid-grafted-magnetite nanocellulose composite. *Journal of Colloid and Interface Science, 456*, 22–31.
5. Ashworth, J., Papps, I., & Storey, D., (1987). Assessing the impact upon the British chlor-alkali industry of the EEC directive on discharges of mercury into waterways. *Land Economics, 63*(1), 72–78.
6. Bassi, A., & Yudken, J., (2009). Potential challenges faced by the US chemicals industry under a carbon policy. *Sustainability, 1*(3), 592–611.
7. Bissett, D. W., & McBeath, I. T., (1988). An overview of the effectiveness of Canadian chlor-alkali mercury regulations. *Toxic Contamination in Large Lakes, 1*, 315–342.
8. Bosch, C., Olivares, A., Faria, M., Navas, J. M., del Olmo, I., Grimalt, J. O., & Barata, C., (2009). Identification of water soluble and particle bound compounds causing sublethal toxic effects. A field study on sediments affected by a chlor-alkali industry. *Aquatic Toxicology, 94*(1), 16–27.
9. Bravo, A. G., Loizeau, J. L., Bouchet, S., Richard, A., Rubin, J. F., Ungureanu, V. G., & Dominik, J., (2010). Mercury human exposure through fish consumption in a reservoir contaminated by a chlor-alkali plant: Babeni reservoir (Romania). *Environmental Science and Pollution Research, 17*, 1422–1432.
10. Bravo, A. G., Loizeau, J. L., Dranguet, P., Makri, S., Björn, E., Ungureanu, V. G., & Cosio, C., (2016). Persistent Hg contamination and occurrence of Hg-methylating transcript (hgcA) downstream of a chlor-alkali plant in the Olt River (Romania). *Environmental Science and Pollution Research, 23*, 10529–10541.

11. Breedveld, L., (2000). A Dutch approach to the European directive on integrated pollution prevention and control: Using life-cycle assessment for the integrated assessment of technologies. *Journal of Industrial Ecology, 4*(2), 35–48.
12. Brinkmann, T., Santonja, G. G., Schorcht, F., Roudier, S., & Sancho, L. D., (2014). *Best Available Techniques (BAT) Reference Document for the Production of Chlor-Alkali*. JRC (Joint Research Center EU), JRC91156.
13. Busto, Y., Cabrera, X., Tack, F. M. G., & Verloo, M. G., (2011). Potential of thermal treatment for decontamination of mercury containing wastes from chlor-alkali industry. *Journal of Hazardous Materials, 186*(1), 114–118.
14. Campanari, S., Guandalini, G., Coolegem, J., Ten Have, J., Hayes, P., & Pichel, A. H., (2019). Modeling, development, and testing of a 2 MW polymeric electrolyte membrane fuel cell plant fueled with hydrogen from a chlor-alkali industry. *Journal of Electrochemical Energy Conversion and Storage, 16*(4), 041001.
15. Chen, D., & Murata, T., (2017). Risk management of chlor-alkali industry based on CVaR. In: *Proceedings of the International Multi-Conference of Engineers and Computer Scient*ists.
16. Crook, J., & Mousavi, A., (2016). The chlor-alkali process: A review of history and pollution. *Environmental Forensics, 17*(3), 211–217.
17. Emmott, N., & Slayne, M., (2005). Legislation and policy concerning mercury in the European Union. In: *Dynamics of Mercury Pollution on Regional and Global Scales: Atmospheric Processes and Human Exposures Around the World* (pp. 65–80). Boston, MA: Springer US.
18. Erdmann, G., & Sartorius, C., (2005). Mercury-based chlor-alkali electrolysis-retrofitting and phase-out as strategic alternatives. *Time Strategies, Innovation, and Environmental Policy*, 79.
19. Gadelha, J. R., Ferreira, V. A. M., Abreu, S. N., Soars, A. M. V. M., Morgado, F. M. R., Hamamura, N., & Tanabe, S., (2010). Experimental mercury bioaccumulation trends in sea anemone *Actinia equina* exposed to chlor-alkali industry effluent contaminated water. *Interdisciplinary Studies on Environmental Chemistry-Biological Responses to Chemical Contaminants: From Molecular to Community Level* (Vol. 3).
20. Garcia-Herrero, I., Margallo, M., Laso, J., Onandía, R., Irabien, A., & Aldaco, R., (2017). Measuring the vulnerability of an energy intensive sector to the EU ETS under a life cycle approach: The case of the chlor-alkali industry. *Sustainability, 9*(5), 837.
21. Garcia-Herrero, I., Margallo, M., Onandía, R., Aldaco, R., & Irabien, A.,

(2017). Environmental challenges of the chlor-alkali production: Seeking answers from a life cycle approach. *Science of the Total Environment, 580*, 147–157.

22. Garcia-Herrero, I., Margallo, M., Onandía, R., Aldaco, R., & Irabien, A., (2017). Life cycle assessment model for the chlor-alkali process: A comprehensive review of resources and available technologies. *Sustainable Production and Consumption, 12*, 44–58.
23. Garcia-Herrero, I., Margallo, M., Onandía, R., Aldaco, R., & Irabien, A., (2018). Connecting wastes to resources for clean technologies in the chlor-alkali industry: A life cycle approach. *Clean Technologies and Environmental Policy, 20*, 229–242.
24. Garron, C., Gagné, F., Ernst, W., Julien, G., Bernier, M., & Caldwell, C., (2005). Mercury contamination of marine sediments and blue mussels (*Mytilus edulis*) in the vicinity of a mercury cell chlor-alkali plant in Dalhousie, New Brunswick, Canada. *Water Quality Research Journal, 40*(1), 1–15.
25. Gibičar, D., Horvat, M., Logar, M., Fajon, V., Falnoga, I., Ferrara, R., & Pacyna, J., (2009). Human exposure to mercury in the vicinity of chlor-alkali plant. *Environmental Research, 109*(4), 355–367.
26. Gibičar, D., Horvat, M., Logar, M., Fajon, V., Falnoga, I., Ferrara, R., & Pacyna, J., (2009). Human exposure to mercury in the vicinity of chlor-alkali plant. *Environmental Research, 109*(4), 355–367.
27. Goodman, D., (1994). Global markets for chlorine and PVC: Potential impacts of Greenpeace attacks. *Journal of Vinyl Technology, 16*(3), 156–161.
28. Grönlund, R., Sjöholm, M., Weibring, P., Edner, H., & Svanberg, S., (2005). Elemental mercury emissions from chlor-alkali plants measured by lidar techniques. *Atmospheric Environment, 39*(39), 7474–7480.
29. Guandalini, G., Foresti, S., Campanari, S., Coolegem, J., & Ten Have, J., (2017). Simulation of a 2 MW PEM fuel cell plant for hydrogen recovery from chlor-alkali industry. *Energy Procedia, 105*, 1839–1846.
30. Guo, J. H., & Sun, W. Y., (2020). Integrating nickel-nitrogen doped carbon catalyzed CO_2 electroreduction with chlor-alkali process for CO, Cl_2, and $KHCO_3$ production with enhanced techno-economics. *Applied Catalysis B: Environmental, 275*, 119154.
31. Hackl, A. E., Sammer, G., & Winter, B., (2001). UN-ECE task force. *Management on By-Products/Residues Containing Heavy Metals and/or Persistent Organic Pollutants*. Status Report (No. CP-031). Federal Environment Agency (Austria).

32. Harris, M., (2008). Phase-out issues for mercury cell technology in the chlor-alkali industry. *Modern Chlor-Alkali Technology, 8*, 19.
33. Hoffmann, C., Weigert, J., Esche, E., & Repke, J. U., (2020). Towards demand-side management of the chlor-alkali electrolysis: Dynamic, pressure-driven modeling and model validation of the 1, 2-dichloroethane synthesis. *Chemical Engineering Science, 214*, 115358.
34. Hofmann, M., Müller, R., Christidis, A., Fischer, P., Klaucke, F., Vomberg, S., & Tsatsaronis, G., (2022). Flexible and economical operation of chlor-alkali process with subsequent polyvinyl chloride production. *AIChE Journal, 68*(1), e17480.
35. Hosseini, M. G., & Zardari, P., (2017). Oxygen depolarized cathode in advanced chlor-alkali cell with Pt-Ru nanoparticles as electro-catalyst: Effect of process conditions and response surface methodology. *Desalination and Water Treatment, 84*, 299–308.
36. Hurskainen, M., & Ihonen, J., (2020). Techno-economic feasibility of road transport of hydrogen using liquid organic hydrogen carriers. *International Journal of Hydrogen Energy, 45*(56), 32098–32112.
37. Jia, X., Xin, G., Qian, Y., & Qian, Y., (2016). Sectoral co-control of air pollutants: Case of a chlor-alkali/polyvinyl chloride sector in China. *Journal of Cleaner Production, 112*, 1667–1675.
38. Jung, J., Postels, S., & Bardow, A., (2014). Cleaner chlorine production using oxygen depolarized cathodes? A life cycle assessment. *Journal of Cleaner Production, 80*, 46–56.
39. Khasawneh, H., Saidan, M. N., & Al-Addous, M., (2019). Utilization of hydrogen as clean energy resource in chlor-alkali process. *Energy Exploration and Exploitation, 37*(3), 1053–1072.
40. Kinsey, J. S., Anscombe, F. R., Lindberg, S. E., & Southworth, G. R., (2004). Characterization of the fugitive mercury emissions at a chlor-alkali plant: overall study design. *Atmospheric Environment, 38*(4), 633–641.
41. Kirk, E., & Silfverberg, H., (2006). Harmonization in the Baltic sea region. *The International Journal of Marine and Coastal Law, 21*(2), 235–258.
42. Klaucke, F., Müller, R., Hofmann, M., Weigert, J., Fischer, P., Vomberg, S., & Repke, J. U., (2023). chlor-alkali process with subsequent polyvinyl chloride production – Cost analysis and economic evaluation of demand response. *Industrial and Engineering Chemistry Research*.
43. Kolon, K., Ruczakowska, A., Samecka-Cymerman, A., & Kempers, A. J., (2015). *Brachythecium rutabulum* and *Betula pendula* as bioindicators of heavy metal pollution around a chlor-alkali plant in Poland. *Ecological Indicators, 52*, 404–410.

44. Kumar, M., & Singh, K., (2022). Fuzzy fault tree analysis of chlorine gas release hazard in chlor-alkali industry using α-cut interval-based similarity aggregation method. *Applied Soft Computing, 125*, 109199.

45. Lakshmanan, S., & Murugesan, T., (2014). The chlor-alkali process: Work in progress. *Clean Technologies and Environmental Policy, 16*, 225–234.

46. Landis, M. S., Keeler, G. J., Al-Wali, K. I., & Stevens, R. K., (2004). Divalent inorganic reactive gaseous mercury emissions from a mercury cell chlor-alkali plant and its impact on near-field atmospheric dry deposition. *Atmospheric Environment, 38*(4), 613–622.

47. Le Faucheur, S., Vasiliu, D., Catianis, I., Zazu, M., Dranguet, P., Beauvais-Flück, R., & Slaveykova, V. I., (2016). Environmental quality assessment of reservoirs impacted by Hg from chlor-alkali technologies: Case study of a recovery. *Environmental Science and Pollution Research, 23*, 22542–22553.

48. Leddy, J. J., (1980). The chlor-alkali industry. *Journal of Chemical Education, 57*(9), 640.

49. Lee, D. Y., Elgowainy, A., & Dai, Q., (2018). Life cycle greenhouse gas emissions of hydrogen fuel production from chlor-alkali processes in the United States. *Applied Energy, 217*, 467–479.

50. Lee, D. Y., Elgowainy, A., & Dai, Q., (2018). Life cycle greenhouse gas emissions of hydrogen fuel production from chlor-alkali processes in the United States. *Applied Energy, 217*, 467–479.

51. Li, C., Zhu, L., & Fleiter, T., (2014). Energy efficiency potentials in the chlor-alkali sector: A case study of Shandong province in China. *Energy and Environment, 25*(3–4), 661–686.

52. Liang, Y., Wang, H., Diao, P., Chang, W., Hong, G., Li, Y., & Dai, H., (2012). Oxygen reduction electrocatalyst based on strongly coupled cobalt oxide nanocrystals and carbon nanotubes. *Journal of the American Chemical Society, 134*(38), 15849–15857.

53. Lindström, M., Attila, M., Ihalainen, T., Kohl, T., Pennanen, J., Sahivirta, E., & Secci, D., (2003). *Energy Efficiency in Environmental Permits*.

54. Liu, G., Zheng, M., Du, B., Liu, W., Zhang, B., & Xiao, K., (2010). Emission of polychlorinated naphthalenes during thermal related processes. In: *EGU General Assembly Conference Abstracts* (p. 2019).

55. Margallo, M., Onandía, R., Aldaco, R., & Irabien, A., (2016). When life cycle thinking is necessary for decision making: Emerging cleaner technologies in the chlor-alkali industry. *Chemical Engineering Transactions, 52*, 475–480.

56. Masnicki, J., (2018). Decentralized application of the BAT conclusions. *European Energy and Environmental Law Review, 27*(2).
57. McHugh, B., Berbee, R., Farmer, E., Fryer, R., Green, N., Larsen, M. M., & ICES Marine Chemistry Working Group, (2016). *Mercury Assessment in the Marine Environment: Assessment Criteria Comparison (EAC/EQS) for Mercury*. OSPAR Commission.
58. Mihaiescu, T., Mihaiescu, R., & Odagiu, A., (2012). Environmental issues within the chlor-alkali manufacturing industry-mercury cell process. *Bull. Univ. Agric. Sci. Vet. Med., 69*, 168–176.
59. Moussallem, I., Jörissen, J., Kunz, U., Pinnow, S., & Turek, T., (2008). Chlor-alkali electrolysis with oxygen depolarized cathodes: history, present status, and future prospects. *Journal of Applied Electrochemistry, 38*, 1177–1194.
60. Mukherjee, A. B., Zevenhoven, R., Brodersen, J., Hylander, L. D., & Bhattacharya, P., (2004). Mercury in waste in the European Union: Sources, disposal methods and risks. *Resources, Conservation, and Recycling, 42*(2), 155–182.
61. Ng, D. K., Chew, I. M., Tan, R. R., Foo, D. C., Ooi, M. B., & El-Halwagi, M. M., (2014). RCNet: An optimization software for the synthesis of resource conservation networks. *Process Safety and Environmental Protection, 92*(6), 917–928.
62. O'Brien, T. F., Bommaraju, T. V., Hine, F., O'Brien, T. F., Bommaraju, T. V., & Hine, F., (2005). History of the chlor-alkali industry. *Handbook of Chlor-Alkali Technology: Volume I: Fundamentals, Volume II: Brine Treatment and Cell Operation, Volume III: Facility Design and Product Handling, Volume IV: Plant Commissioning and Support Systems, Volume V: Corrosion, Environmental Issues, and Future Development*, 17–36.
63. O'Brien, T. F., Bommaraju, T. V., Hine, F., O'Brien, T. F., Bommaraju, T. V., & Hine, F., (2005). Overview of the chlor-alkali industry. *Handbook of Chlor-Alkali Technology: Volume I: Fundamentals, Volume II: Brine Treatment and Cell Operation, Volume III: Facility Design and Product Handling, Volume IV: Plant Commissioning and Support Systems, Volume V: Corrosion, Environmental Issues, and Future Development*, 37–74.
64. Otashu, J. I., & Baldea, M., (2019). Demand response-oriented dynamic modeling and operational optimization of membrane-based chlor-alkali plants. *Computers and Chemical Engineering, 121*, 396–408.
65. Pacyna, E. G., Pacyna, J. M., Fudala, J., Strzelecka-Jastrzab, E., Hlawiczka, S., & Panasiuk, D., (2006). Mercury emissions to the atmosphere from anthropogenic sources in Europe in 2000 and their scenarios until 2020.

Science of the Total Environment, 370(1), 147–156.

66. Park, I. K., Ahn, C. Y., Lee, J. H., Lee, D. W., Lee, C. H., Cho, Y. H., & Sung, Y. E., (2019). Three-dimensionally interconnected titanium foam anode for an energy-efficient zero gap-type chlor-alkali electrolyzer. *International Journal of Hydrogen Energy, 44*(31), 16079–16086.
67. Raldúa, D., Díez, S., Bayona, J. M., & Barceló, D., (2007). Mercury levels and liver pathology in feral fish living in the vicinity of a mercury cell chlor-alkali factory. *Chemosphere, 66*(7), 1217–1225.
68. Rallo, M., Lopez-Anton, M. A., Contreras, M. L., & Maroto-Valer, M. M., (2012). Mercury policy and regulations for coal-fired power plants. *Environmental Science and Pollution Research, 19*, 1084–1096.
69. Reis, A. T., Rodrigues, S. M., Araújo, C., Coelho, J. P., Pereira, E., & Duarte, A. C., (2009). Mercury contamination in the vicinity of a chlor-alkali plant and potential risks to local population. *Science of the Total Environment, 407*(8), 2689–2700.
70. Santoro, A., Terzano, R., Blo, G., Fiore, S., Mangold, S., & Ruggiero, P., (2010). Mercury speciation in the colloidal fraction of a soil polluted by a chlor-alkali plant: A case study in the South of Italy. *Journal of Synchrotron Radiation, 17*(2), 187–192.
71. Scheme, E. E. T., (2010). *The European Chlor-Alkali Industry: An Electricity Intensive Sector Exposed to Carbon Leakage*. Brussels: Euro Chlor.
72. Scoullos, M. J., Vonkeman, G. H., Thornton, I., Makuch, Z., Scoullos, M. J., Vonkeman, G. H., & Makuch, Z., (2001). Policy options for mercury. *Mercury–Cadmium–Lead Handbook for Sustainable Heavy Metals Policy and Regulation*, 409–425.
73. Shen, Z., Zhao, H., Liu, Y., Kan, Z., Xing, P., Zhong, J., & Jiang, B., (2018). Mercury-free nitrogen-doped activated carbon catalyst: An efficient catalyst for the catalytic coupling reaction of acetylene and ethylene dichloride to synthesize the vinyl chloride monomer. *Reaction Chemistry and Engineering, 3*(1), 34–40.
74. Shojaikaveh, N., Ashrafizadeh, S. N., Mohammadi, F., & Amerighasrodashti, A., (2010). Optimization of predicted cell voltage and caustic current efficiency in a chlor-alkali membrane cell with application of genetic algorithm. *ECS Transactions, 25*(33), 43.
75. Southworth, G. R., Lindberg, S. E., Zhang, H., & Anscombe, F. R., (2004). Fugitive mercury emissions from a chlor-alkali factory: Sources and fluxes to the atmosphere. *Atmospheric Environment, 38*(4), 597–611.
76. Ullrich, S. M., Ilyushchenko, M. A., Kamberov, I. M., & Tanton, T. W.,

(2007). Mercury contamination in the vicinity of a derelict chlor-alkali plant. Part I: Sediment and water contamination of Lake Balkyldak and the River Irtysh. *Science of the Total Environment, 381*(1–3), 1–16.

77. Walker, T. R., (2016). Mercury concentrations in marine sediments near a former mercury cell chlor-alkali plant in eastern Canada. *Marine Pollution Bulletin, 107*(1), 398–401.
78. Wang, X., Teichgraeber, H., Palazoglu, A., & El-Farra, N. H., (2014). An economic receding horizon optimization approach for energy management in the chlor-alkali process with hybrid renewable energy generation. *Journal of Process Control, 24*(8), 1318–1327.
79. Wängberg, I., Edner, H., Ferrara, R., Lanzillotta, E., Munthe, J., Sommar, J., & Weibring, P., (2003). Atmospheric mercury near a chlor-alkali plant in Sweden. *Science of the Total Environment, 304*(1–3), 29–41.
80. Whitfield, R., & Brown, F., (2014). Climate change, terrorism concerns, and technology choice in the water industry. *The International Journal of Sustainability Policy and Practice, 8*(4), 97.
81. Yarime, M., (2007). Promoting green innovation or prolonging the existing technology. *Journal of Industrial Ecology, 11*(4), 117–139.
82. Yumashev, A. V., Fateminasab, S. M., Marjani, A., & Lirgeshas, A. B., (2021). Development of computational methods for estimation of current efficiency and cell voltage in a chlor-alkali membrane cell. *Energy Sources, Part A: Recovery, Utilization, and Environmental Effects*, 1–14.
83. Zhang, L. N., Lang, Z. L., Wang, Y. H., Tan, H. Q., Zang, H. Y., Kang, Z. H., & Li, Y. G., (2019). Cable-like Ru/WNO@ C nanowires for simultaneous high-efficiency hydrogen evolution and low-energy consumption chlor-alkali electrolysis. *Energy and Environmental Science, 12*(8), 2569–2580.
84. Zhang, L. N., Li, R., Zang, H. Y., Tan, H. Q., Kang, Z. H., Wang, Y. H., & Li, Y. G., (2021). Advanced hydrogen evolution electrocatalysts promising sustainable hydrogen and chlor-alkali co-production. *Energy and Environmental Science, 14*(12), 6191–6210.
85. Zhang, X., Wang, Y., Wei, S., Dong, J., Zhao, J., & Qian, G., (2022). Assessing the chlorine metabolism and its resource efficiency in chlor-alkali industrial symbiosis: A case of Shanghai Chemical Industry Park. *Journal of Cleaner Production, 380*, 134934.
86. Zhou, Z., Liu, Y., & Jin, C., (2009). Study on construction of low-carbon featured chlor-alkali industry bases energized by non-grid-connected wind power in offshore Jiangsu province. In: *2009 World Non-Grid-Connected Wind Power and Energy Conference* (pp. 1–4). IEEE.

CHAPTER

8 Cost Saving in Chlorine Plants

Contents

8.1. INTRODUCTION

This chapter will provide a concise overview of the origin and price of titanium. Despite not being classified as a precious or rare metal, titanium is widely regarded as a costly industrial metal. This particular metallic element ranks fourth in terms of abundance and ninth in terms of overall abundance among all elements found in the earth's outer layer. The commercially valuable oxide ore of this element is found in minerals such as rutile and ilmenite, as well as various iron ores. There are abundant exploitable deposits of this ore located worldwide, including in Australia, India, Canada, Russia, Norway, Malaysia, South Africa, and the United States. Owing to its distinctive characteristics and elevated expense, it has frequently been denoted as unobtanium. Titanium is considered to be the most recently developed among the primary structural metals. In contrast to iron, which has been documented to have been utilized by humans as early as 3200 BC, titanium wasn't identified as a distinct element until the latter part of the 18th century. The commercial production of titanium was not demonstrated until 1937, when Dr. Wilhelm J. Kroll, an inventor from Luxembourg, developed a process for its production. The Kroll process entails the reduction of titanium tetrachloride through a chemical reaction with magnesium. The metal is typically obtained by subjecting the sponge-like material to either acid leaching or more usually vacuum distillation processes, which serve to eliminate any impurities present. Following Kroll's demonstration, the U.S. Bureau of Mines invested an additional 11 years in the development process before successfully producing the initial commercially available titanium sheet. This chapter aims to provide a definition of the term 'unique properties' and analyze how these properties are utilized in various industrial processes such as the membrane, diaphragm, and mercury cell processes. Additionally, this chapter will briefly discuss other miscellaneous applications of these unique properties. The final section of the chapter will provide a synopsis of the monetary advantages that drive the utilization of this complex material (Tak & Kumar, 2017; Ullman, 2008).

8.2. PRODUCTION OF TITANIUM

The production of titanium metal involves a three-step process, which includes the transformation of ore into a mill product (Fray, 2008):

- The process of chemically reducing ore to form sponge involves the formation of agglomerated granules that bear a resemblance to a sea sponge (Perego et al., 2001).

- The process of forming an ingot involves melting the sponge, which is often done together with titanium scrap (Hu et al., 2018).
- Finally, the last step involves the transformation of the ingot into marketable mill products (Henriques et al., 2005).

The metallurgical procedure involved in the conversion of ore into metallic sponge is a complex, multi-stage, high-temperature batch process that demands significant investments of labor, energy, and capital. Despite efforts to enhance it, the procedure has essentially remained unchanged since its origination (Panigrahi et al., 2010).

The process of transforming sponge into ingot is a complex one, requiring significant amounts of energy regardless of the specific melting technique employed, whether it be vacuum arc re-melting (VAR), electron beam cold hearth melting (EBCHM), and plasma arc melting (PAM). Similar to the chemical reduction process of ores, the reduction process of metals must be carried out under vacuum or inert atmosphere to regulate reactive contamination that could potentially jeopardize the metal's strength (Araci et al., 2015).

The final stage, involving the thermo-mechanical transformation of titanium ingot to various mill products such as bars, plates, sheets, rods, etc., is executed in a manner that is similar to that of other metals. However, the significant reactivity of the metal is of utmost importance. The metal undergoes heating to attain the suitable temperature, followed by processing to achieve the subsequent incremental size or shape, primarily through forging or rolling techniques. Subsequently, it is permitted to cool down, conditioned, and subjected to inspection. Subsequently, the aforementioned procedure is reiterated until the ultimate form and dimensions of the mill are attained. Upon exposure to elevated temperatures in the presence of air, the metal exhibits reactivity by undergoing absorption of oxygen and nitrogen. This results in the formation of a rigid and fragile oxygen-enriched stage of the metal, commonly referred to as "alpha case," which encompasses the entirety of its surface.

The mechanical characteristics of the alpha case layer exhibit a significant reduction in comparison to those of the parent metal. Furthermore, during the cooling process of the metal, surface cracks arise that have the potential to penetrate the material up to a thickness of 5% or beyond. These cracks are also encompassed by alpha case. The removal of the alpha case layer and cracks is imperative prior to any further thermo-mechanical processing. Failure to do so may result in the exacerbation of cracks and defects, leading to compromised performance as well as fatigue properties of the metal. This, in turn, may result in significant yield losses during subsequent conditioning steps (Poletti et al., 2008).

8.2.1. Origin and Cost of Titanium

Titanium is a material that is not considered rare and is classified as the fourth most abundant element in the crust of the earth. Rutile deposits are widely distributed across the globe, and the majority of refined titanium dioxide is utilized in pigments. Its exceptional stability renders it suitable for a wide range of uses in the paint as well as paper sectors. A minute proportion of titanium dioxide that undergoes conversion to metal undergoes a series of intricate procedures with the objective of generating titanium of exceptional purity in the shape of ingots, slabs, and other forms that can be subsequently refined into semis including tubes, sheets, bars, and the like. In 1999, the overall yearly output of titanium was 50,000 tons, out of which 24,000 tons were subjected to alloying with aluminum, vanadium, and other metals. This was done to augment the strength and fatigue characteristics of the titanium alloy, specifically for its application in the aerospace sector. Titanium, which ranks ninth in terms of abundance among the elements present in the earth's crust, occurs naturally in diverse oxide forms, with titanium dioxide being the most prevalent. Nonetheless, the aforementioned substance is not encountered in an unadulterated form, but rather is invariably intermingled with other minerals or exists as a constituent of intricate minerals that commonly comprise iron, manganese, magnesium, as well as other elements' oxides. The minerals containing titanium are not readily utilizable by end-users and require a sequence of metallurgical as well as chemical procedures, which entail the application of chlorine chemistry, to extract and refine the titanium (Shi et al., 2023; Livraghi et al., 2006).

Titanium possesses a unique combination of properties such as low density, high strength, and superior corrosion resistance compared to a wide range of nonferrous and ferrous metals as well as alloys. The metal in question exhibits the greatest strength-to-weight ratio among all metals, with tensile strengths that are akin to those of steel, while its density is approximately 45% less. The material's elevated melting point, robustness under high temperatures, and ability to resist oxidation render it highly appropriate for deployment in corrosive, high-temperature settings, such as those encountered by aerospace components as well as combustion turbines. The corrosion-resistant properties of titanium render it a valuable material for deployment in highly challenging environments, such as chemical, metallurgical, as well as power plants, where exposure to acidic as well as saline solutions is commonplace. Minor quantities of this substance are utilized as an alloying element in various alloys and metals and are applied in diverse consumer products and health-related apparatuses such as pacemakers, artificial hips, and knees (Ma et al., 2006; Hummer et al., 2009).

The 26,000-ton equilibrium was primarily subjected to sheet processing and employed in the chemical sector and other domains where the value of uptime is significant enough to warrant the replacement of conventional materials with titanium (Clerici, 2015).

The prevailing market value of unalloyed titanium is estimated to be within the range of US $15–20 k This is attributed to its lower specific weight in comparison to stainless steel, resulting in a relative cost per meter of pipe ranging from US $9–12 20 k. It is noteworthy to consider this aspect when making a comparison about high-molybdenum stainless steels, which are currently valued at approximately US $9 20 k (Hu et al., 2020).

8.3. UNIQUE PROPERTIES

Titanium exhibits several distinctive characteristics. The present chapter will center on the resistance to corrosion of the metal, that is closely linked to one of the distinctive characteristics of titanium, namely, its enduring affinity for oxygen (Ohno et al., 1997).

The strong intermolecular attraction between the two chemical species yields a malleable, robust, and autoregenerative film of oxide, which confers upon titanium its distinctive functional properties. Titanium is a chemically reactive metallic element that, in the absence of its naturally occurring oxide layer, would undergo dissolution in aqueous environments (Setoyama et al., 2004).

Titanium possesses distinctive design characteristics, namely:

- Corrosion resistance;
- Light weight;
- Erosion resistance;
- High specific strength;
- Low thermal expansion;
- Good heat transfer.

Titanium is utilized in the chlor-alkali production as a material for activated metal anodes due to its distinctive properties. The predominant application of titanium lies in its utilization within the chlor-alkali sector (Boyer, 2010).

The utilization of titanium is observed in the Krupp Uhde membrane cell anode, as depicted in Figure 8.1. The phenomenon of anodic current flow in titanium occurs only after the breakdown voltage threshold has been surpassed. The voltage observed in typical brine under natural pH conditions is within the range of 7 to 14 V. The specific value is contingent upon factors such as the metal's level of purity, the standard of the brine, and other relevant variables. The implementation of a high-voltage obstacles in commercial applications would be deemed impractical. However, this obstacle can be surmounted by amalgamating titanium oxide alongside a noble metal oxide that possesses an identical crystal structure. This facilitates the merging of oxides, resulting in a genuinely mixed oxide. This composite material consisting of two or more oxides (Khorasani et al., 2015; Kang et al., 2019).

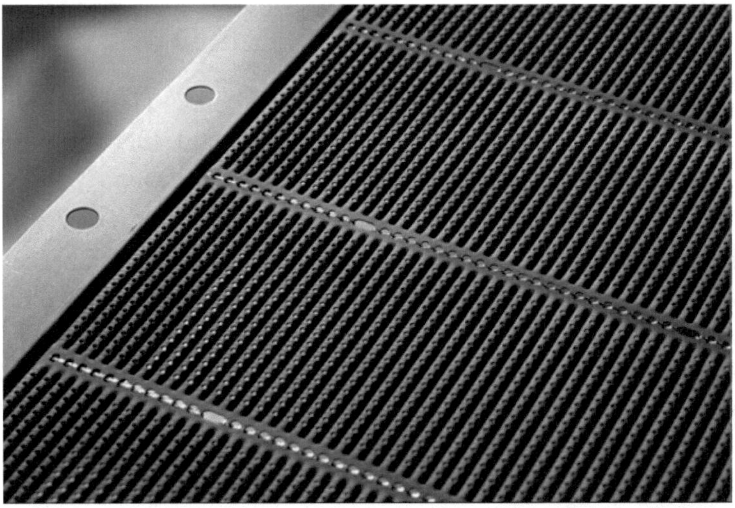

Figure 8.1. *Illustration of Krupp Uhde membrane cell anode using titanium.*

Source: A. Ullman, Creative Commons License.

The mixed oxide in question serves as the Beer covering for the electrode within the electrolysis cell and was first discovered during the latter half of the 1960s (Turkevych et al., 2008).

The Beer coating, marketed as DSA, possesses a notable benefit in its exceptional affinity to the underlying electrode substrate, owing to the presence of titanium oxide within the coating. Contemporary DSA coatings bear minimal resemblance to the initial ruthenium Beer coatings (Notari, 1988).

The following analysis pertains to Figures 8.2–8.7, which are derived from the research conducted by the esteemed French chemist, Pourbaix. These figures illustrate the correlation between voltage and pH for various metals (Fan et al., 2019).

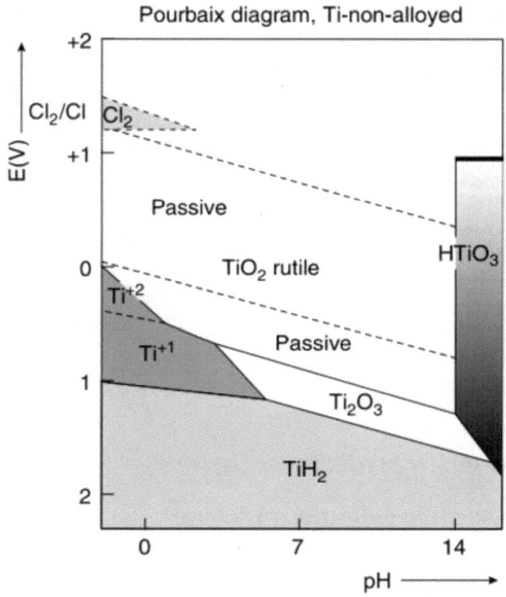

Figure 8.2. *Pourbaix diagram for non-alloyed titanium.*

Source: John Moor house, Creative Commons License.

Under a wide range of potentials and pH values, titanium exhibits passivity and does not undergo corrosion. The white regions depicted in Figure 8.2 unveil this information. Nevertheless, there exist certain exemptions. The phenomenon of alkaline attack should be taken into consideration (as depicted in Figure 8.3). There exists a common assumption that the utilization of titanium is not feasible under alkaline conditions such as hypochlorite and excessive alkalinity. The assumption cannot be supported by the available corrosion data. Corrosion and erosion are significant concerns only at elevated temperatures and concentrations (Hantanasirisakul et al., 2019).

An additional instance that challenges conventional knowledge pertains to the response of titanium when exposed to reducing conditions at low pH levels. The flange located in an anolyte line (as shown in Figure 8.4) should be taken into consideration (Parrino et al., 2021).

- with hot brine (90°C);
- with saturated with Cl_2;
- a high Eredox;
- a low pH.

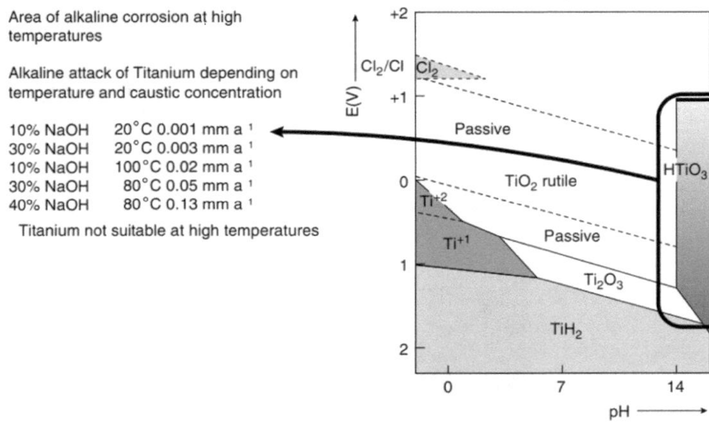

Figure 8.3. *Representation of alkaline corrosion area at high temperatures.*

Source: M. Harris, Creative Commons License.

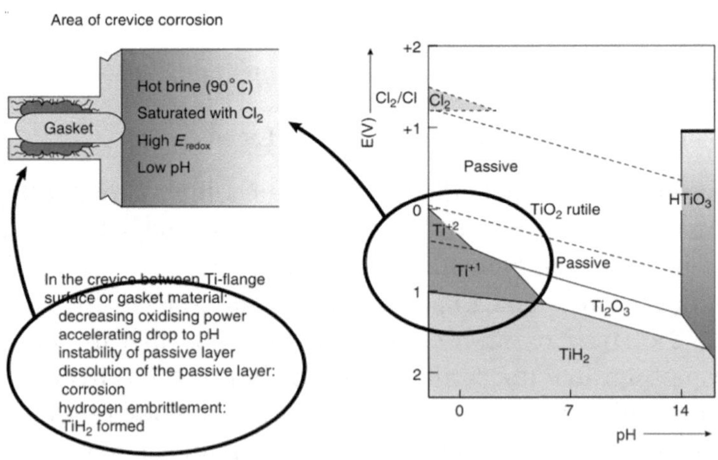

Figure 8.4. *Area of crevice corrosion.*

Source: T. Kimura, Creative Commons License.

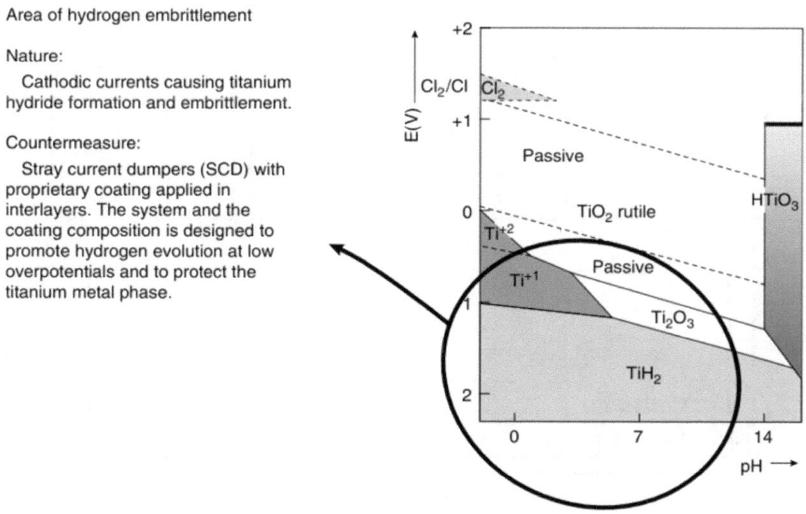

Figure 8.5. *Illustration of area of hydrogen embrittlement.*

Source: T. Kimura, Creative Commons License.

The utilization of proprietary coating in interlayers for stray current dumpers (SCD) is being employed. The coating composition and system have been developed with the objective of facilitating hydrogen evolution at minimal over potentials while simultaneously safeguarding the titanium metal stage.

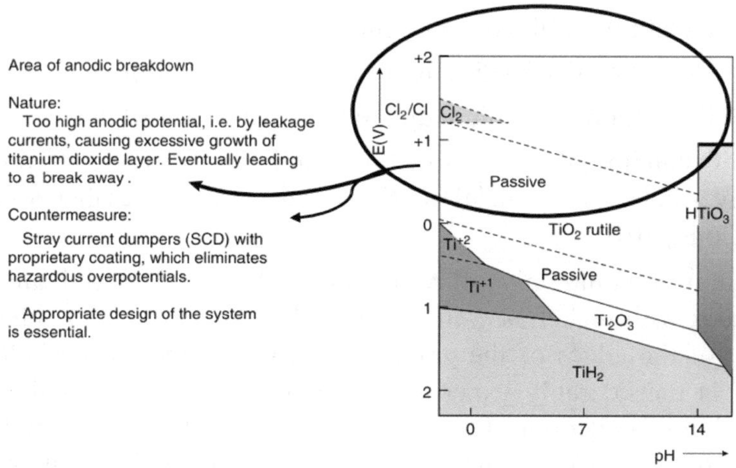

Figure 8.6. *Representation of area of anodic breakdown.*

Source: John Moorhouse, Creative Commons License.

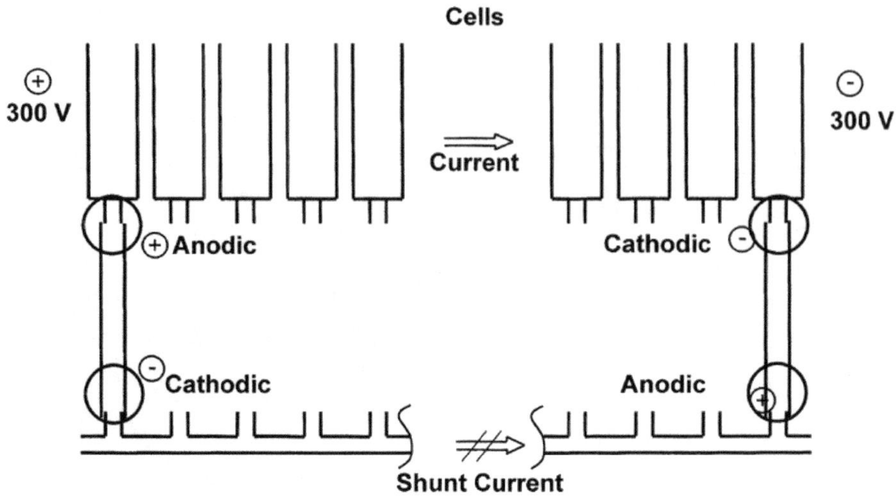

Figure 8.7. *Illustration of shunt currents, local anodes and cathodes due to electrolysis and cell design.*

Source: A. Ullman, Creative Commons License.

It is plausible to come across certain elements within the crevice situated within the Ti flange surface and gasket material:

- Reduction in oxidizing potential;
- Rapid decrease in pH;
- Instability of the passive film;
- Corrosion due to dissolution of the passive film;
- Formation of titanium due to hydrogen embrittlement.

Notwithstanding, there exist methods and strategies to overcome these issues; nevertheless, it is crucial to prioritize design features that ensure high quality (Dong, 2010).

Titanium is commonly alloyed with palladium and ruthenium, and occasionally with nickel as well as molybdenum, in order to enhance the durability of the edges of the grains within the flange. Nevertheless, this approach is considerably expensive, particularly considering the current market value of palladium. Permascand has devised a commercial system for noble coatings that are administered through electrochemical or thermal means. These coatings have been employed for over 20 years and exhibit the capability to withstand routine gasket replacements (Thompson, 2000). The phenomenon of repressed voltage is a typical occurrence in electrochemical facilities, albeit necessitating a certain degree of vigilance. Regarding this matter, it is advisable to take into account the phenomenon of cathodic polarization. Whilst titanium exhibits exceptional performance as an anode, it presents certain limitations when functioning as a cathode. The primary cause of this phenomenon is the creation of substantial amounts of titanium hydride. This substance, in addition to flaking off, causes the cathode material to become fragile, leading to the propagation of cracks (as depicted in Figure 8.5). Prevention of hydride formation is a crucial requirement (Ouchi et al., 1999).

The optimal approach entails developing a cathode that exhibits decreased polarization exposure, thereby affording abbreviated instances of heightened polarization intensity or protracted durations at diminished current densities. The latter can be accommodated, provided that the appropriate protective coating is administered. A group of catalytic coatings can effectively decrease the activation energy at the surface of the cathode, thereby facilitating the escape of hydrogen gas at a significantly reduced overvoltage. This process results in hydride formation at a notably sluggish pace (Gerdemann, 2001).

The management of anodic overvoltage within the catalytic covering is also imperative, as depicted in Figure 8.6. The potential outcome of this phenomenon is the proliferation of titanium dioxide beyond the optimal level, which may ultimately result in system failure. The utilization of titanium stray current dumpers can effectively mitigate the occurrence of excessive potentials.

A representation of a membrane cell header is depicted in Figure 8.7, wherein the battery voltage has the potential to reach a maximum of 600 V. The reduction of voltage can be achieved through the appropriate design of electrodes based on titanium.

Figures 8.2–8.7 serve the purpose of demonstrating that the appropriate design and selective employment of cell units may improve the distinctive characteristics of titanium (Hayat et al., 2019).

8.4. APPLICATIONS

8.4.1. The Membrane Process

Titanium has been extensively utilized in membrane technology for various industrial applications such as anolyte headers, anolyte boxes, filters, exchangers for heat, chlorate removal systems, and different components of the brine system (Chin et al., 2001). Numerous manufacturers of chlorine are compelled to endure impurities that arise from conventional plastic materials employed in membrane technology. The plastic construction materials are subjected to significant stresses from the emergence of hot chlorine and hot anolyte from the anodic portion of the cell (Pan et al., 2012).

The chemical element chlorine exhibits a high degree of reactivity towards the majority of plastic materials, with the exception of those that are composed of polymers possessing a high degree of fluorination. The by-products of chlorine resulting from these reactions exhibit a tendency to not only amalgamate into the chlorine stream yet additionally amass within the plastic material per se. The disposal of contaminated plastic presents challenges in terms of ensuring its safe destruction, particularly given the growing number of restrictions on the introduction of large chlorinated molecules into the environment (Bet-Moushoul et al., 2016).

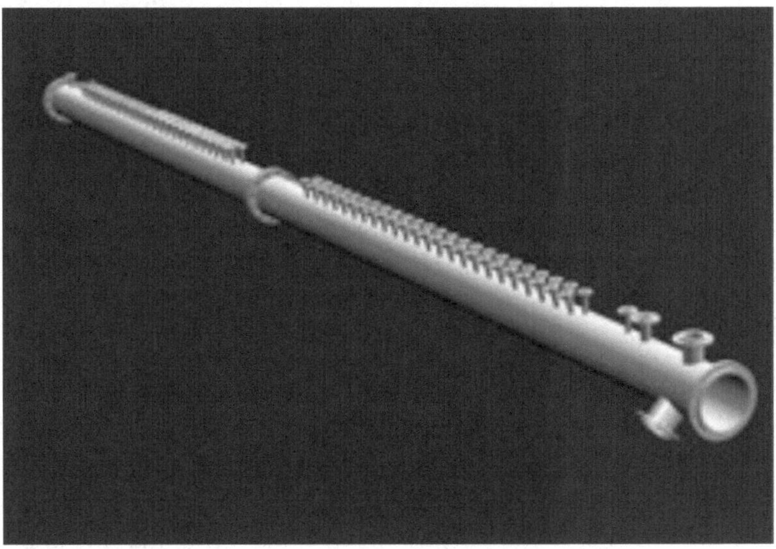

Figure 8.8. *Illustration of a titanium header.*

Source: R. Beckmann, Creative Commons License.

In addition, the replacement of a cell header that has 70–100 connections pose a significant challenge, requiring the incorporation of numerous safety features into its design. The application of non-destructive testing to composite materials is not optimal, and the occurrence of a joint leakage may result in the separation of the chemical-resistant liner via its mechanical support due to delamination. Given the difficulty in detecting leaks under such conditions, it is necessary to replace the entire header during the associated electrolyzer downtime (Guizard, 1996).

A titanium-based system that is appropriately designed with stray current dumpers (as shown in Figure 8.8) is capable of fulfilling the essential functions of the cell header. Furthermore, it is relatively straightforward to conduct repairs to damages (Asif, & Zhang, 2021). Akzo Nobel has been utilizing titanium systems for over 15 years, and to date, no instances of wear have been reported. The fundamental concept underlying the header is to mitigate stray current through the implementation of multiple superfluous stray current dumpers. The diagram depicted in Figure 8.9 illustrates the placement of stray current dumpers situated within the confines of diminutive, flanged nozzles. Additionally, a mesh basket is positioned within the larger pipe, functioning as a secondary safeguard (David et al., 2014).

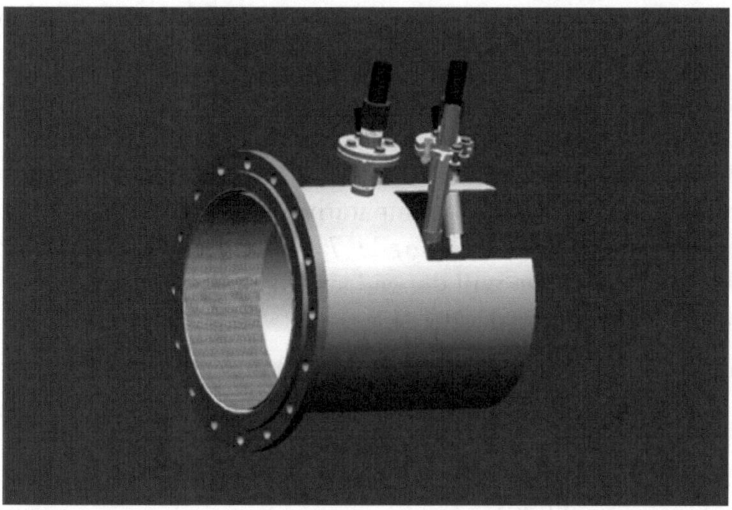

Figure 8.9. *Illustration of a stray current dumper.*

Source: R. Beckmann, Creative Commons License.

There is a growing trend in which rubber-lined filter bodies are being substituted with solid titanium vessels or retrofitted with titanium lining. The rationale behind this measure pertains to the membrane's susceptibility to contaminants, including but not limited to iron. The gradual tightening of impurity sources has led to the utilization of titanium as a separator or frame in retrofitting filter frames during the conversion of cell rooms to membrane technology, which serves as a significant application for this material (Chen & Deng, 2012).

The utilization of lightweight and robust foraminous frame structures, such as titanium mesh, can effectively bolster the filter cloth, resulting in an augmented active surface area of the membrane. Membrane filters have emerged as a contemporary advancement in classical filtering techniques. The size of the particles in the filtrate is determined by the membrane itself. Since its inception, Permascand has been engaged in the utilization of GORE-TEX membranes (Amiri et al., 2022).

In addition to their independence from filtering aids like ¤-cellulose, membrane filters possess a high degree of automation feasibility, as they can be promptly integrated into the system following back-pulsing. Permascand as well as Akzo Nobel have developed registers that are appropriate for retrofitting purposes, which are also utilized by numerous other facilities, including the ones located at Bayer Uerdingen. The performance of these plants' operations has been deemed successful (Abdullah & Kamarudin, 2015; Singh et al., 2021).

The pH level in the chlorate elimination system may exhibit a significant decrease. The favored material in this context is a form of titanium that has been alloyed. The concentration of noble alloy components is observed to be at a level of 0.05%, which is comparatively lower than the concentration present in the well-established grade 7 alloy. The determination of the appropriate alloy to utilize will depend significantly on the temperature and pH tolerance (Razmjou et al., 2012).

Titanium is commonly utilized in membrane cell processes for various purposes such as in the hot anolyte pipelines, anolyte collecting tanks, and dechlorination systems. These systems often operate under vacuum conditions and may restrict the use of unsecured linings (Jiang et al., 2009; Ehsani & Aroujalian, 2020).

8.4.2. The Diaphragm Process

In contrast to the electrolysis methods employed in other processes, the utilization of a diaphragm cell room in the electrolysis of brine does not involve the recirculation of the brine. Additionally, the temperature within the system can be selectively adjusted to achieve optimal conditions. As a result, the quantity of titanium utilized in this process is relatively minimal. Nonetheless, there exist distinct potential options. Permascand has recently patented a novel design for the cell blanket, which incorporates bellows that are joined to the anode collar. This serves as an illustrative instance. Titanium can be utilized in the production of the chlorine header as well as cell top, among other components. The utilization of high-temperature chlorine in combination with FRP materials is not feasible for all end-users. Furthermore, the employment of such consumable materials has become increasingly costly due to regulatory measures. Therefore, there could be potential opportunities for the utilization of titanium alloys (Tan & Miao, 2009; Bewer et al., 1982; Bußmann et al., 2021).

8.4.3. The Mercury Process

It is a widely acknowledged fact among operators of mercury cell rooms that the corrosion of rubber-lined components of the cell is a common phenomenon. The rate of corrosion is directly proportional to the softness of the rubber material. A loose liner has been devised for mercury cells, which is installed on-site in the anode workshop and secured with titanium nuts. The Bakelite pressure plate is utilized to secure the riser tube (Candeago et al., 2022; Noel et al., 2007; Yang et al., 2011).

The application of titanium lining can be executed in two ways: it can be superimposed on an existing rubber lining, which necessitates the use of high-pressure water or a comparable technique for cleaning, or it can be directly applied onto the steel casing. In the latter scenario, certain measures must be implemented to prevent the flow of brine condensates into the interstice among the steel cover as well as the titanium lining (Ghorishi & Sedman, 1998; Economou, & Fielden, 2003; Pelegrini et al., 2001).

A header system was devised to manage the hot chlorine gas emanating from the cells. This system includes stray current dumpers that effectively handle the brine condensates, which could otherwise cause harm (Xie et al., 2012; Gao et al., 2019; Tsai et al., 2019).

8.4.4. Miscellaneous Applications

Titanium is a widely utilized material in various applications where it is crucial to uphold a substantial proportion of tensile strength concerning density. These applications include but are not limited to space and aircraft, automobiles, motorcycles, as well as naval vessels. The utilization of this technology is deemed advantageous due to its ability to enhance fuel efficiency, durability, as well as overall safety of the product. Titanium is utilized in a diverse range of industries, including aerospace, industrial, natural gas and oil, medical, and architectural. Additionally, it is incorporated into various consumer products such as grills, golf clubs, as well as shafts for lacrosse sticks, tennis rackets, as well as framing for bicycles. The material finds application in surgical devices, implants, medical components, crutches, wheelchairs, and many other related areas. The significance of titanium, and any additional material, is contingent upon the application for which it is utilized. When choosing titanium as well as its alloys for a given application, engineers must evaluate the economic as well as technological rationale for employing this material in particular components (Williams & Boyer, 2020; Clark & Roebuck, 1992; Maykuth & Wood, 1969).

Despite the current high cost of titanium, there are numerous civilian and military applications that can be deemed justifiable for the use of this material. In numerous instances, the preliminary elevated expense of the substance is counterbalanced by either the benefits of diminished weight attributable to the metal's low density or by the extended lifespan of the part due to the metal's heightened resistance to corrosion. Given the prevailing expectation that the selling price of this particular metal will likely decline as production levels increase and processing techniques improve, it is not the intention of this discourse to comprehensively address the associated economic factors. The purpose of this analysis is to examine the technological rationale necessary for employing titanium in the desired components (Froes, 2015; Eisenbarth et al., 2004; Moeller, 1950).

It is essential to acknowledge two fundamental factors: the first one is derived from the component's specifications, while the second one is related to the design. Typically, specifications necessitate the fulfillment of specific mechanical properties that are deemed advantageous in the final product. The aforementioned characteristics may encompass singularly or in combination: breaking strength, tensile strength, lengthening, decrease in area, bend ductility, strength of impact, toughness, strength against fatigue, creep, as well as elevated temperature properties. Once the materials engineer has chosen a material that satisfies the minimum requirements, the product engineer assumes the responsibility of evaluating the manufacturing challenges that are specific to the design. The evaluation of the materials' ability to undergo the necessary fabrication techniques to achieve the intended final product is imperative (Kubaschewski et al., 2007; Bhardwaj et al., 2018; Freese et al., 2001).

Aluminum exhibits a favorable strength-to-weight ratio, whereas stainless steel possesses greater strength albeit with a higher weight. Titanium exhibits equivalent strength to stainless steel, while possessing a lower weight. Both stainless steel and titanium exhibit resistance to corrosion. Stainless steel is equipped with a chromium oxide layer, while titanium features a titanium nitride layer, both of which serve as protective barriers against rust. In comparison to aluminum, titanium exhibits a higher weight but possesses twice the strength. Titanium can serve as a suitable substitute for either metal, depending on the specific application (Lakshmanan et al., 2012; Santos, 2017; Khor et al., 2003).

Pure commercial titanium exhibits remarkable stability in high-temperature settings, withstanding temperatures as high as 572°F. This is attributed to its exceptional strength, as well as its notable resistance to thermal and creep deformation. In contrast to aluminum, titanium has the propensity to accumulate heat due to its low thermal and electrical conductivity (Welsch et al., 1993; Ikeda & Shoesmith, 1997; Sioshansi, 1987).

Titanium finds application in the fabrication of anolyte pipes, anolyte containers, dissolver tanks, and maturing tanks within the brine system. In certain cases, the utilization of pure titanium may not be deemed justifiable, thereby necessitating the adoption of alternative materials such as rubber linings and polymeric substances. In situations where there is a stringent requirement for uptime, the utilization of titanium-coated steel could be contemplated as a substitute for rubber (Vieira et al., 2020; Boyer, 1996; Loria, 2000).

A novel lining technique was developed in Sweden to cater to the demand for large hypochlorite tanks. The aforementioned methodology is implemented in the context of extensive reaction vessels, which are fabricated to accommodate volumes ranging from 100 to 150 cubic meters. The utilization of titanium alloy is evident in the lining. In situations where greater levels of strength are necessary or where particular applications mandate the inclusion of specific alloying elements, it becomes imperative to contemplate the utilization of a titanium alloy grade. Elevating the alloy composition tends to enhance the strength up to a certain threshold, albeit at the expense of ductility. The decrease in ductility implies a reduction in the level of malleability. When opting for an alloy, it is typically advantageous to select the alloy that provides the highest level of formability for the desired strength level, regardless of whether the durability requirement pertains to tensile, fatigue, and creep power. In situations where optimal hardness and strength are essential, it may be advantageous to opt for an alloy that demonstrates a favorable reaction to heat-treatment. By subjecting the material to heat treatment, it is possible to achieve optimal ductility, thereby facilitating the process of forming the material into desired shapes.

Consequently, the resultant products can undergo thermal processing to attain the desired level of strength or hardness (Darwish et al., 2011; Pourbaix, 1974; Rust et al., 1990).

Nickel is a prospective material that can be utilized for the construction of cells. As an illustration, Eka Chemicals has utilized nickel cartridges featuring conventional GORE-TEX membranes to eliminate graphite particles from a caustic decomposer in a direct manner. The outcome exhibited favorable indications. In addition to the removal of graphite, the levels of mercury were decreased to less than 10 μg. In addition to the utilization of the cleaner caustic, a significant reduction in sludge volume was observed due to the elimination of the pre-coat requirement in this particular process (Balazic et al., 2007; Elrod, 1980; Li et al., 2000).

8.5. SUMMARY

This chapter presents a concise overview of the economic factors that underlie the utilization of titanium. The savings reported were derived from end-user interviews. This chapter has presented several instances where the financial justification for the utilization of titanium alternatives or original titanium parts is evident. Despite the occasional higher cost, the initial investment in titanium can be deemed justifiable. The enduring nature of product quality persists long after the memory of its price has faded.

REFERENCES

1. Abdullah, N., & Kamarudin, S. K., (2015). Titanium dioxide in fuel cell technology: An overview. *Journal of Power Sources, 278*, 109–118.
2. Amiri, S., Vatanpour, V., Mansourpanah, Y., & Khataee, A., (2022). Recent trends in application of nanoscale zero-valent metals and metals single atoms in membrane processes. *Journal of Environmental Chemical Engineering*, 107457.
3. Araci, K., Mangabhai, D., & Akhtar, K., (2015). Production of titanium by the Armstrong Process®. In: T*itanium Powder Metallurgy* (pp. 149–162). Butterworth-Heinemann.
4. Asif, M. B., & Zhang, Z., (2021). Ceramic membrane technology for water and wastewater treatment: A critical review of performance, full-scale applications, membrane fouling and prospects. *Chemical Engineering Journal, 418*, 129481.
5. Balazic, M., Kopac, J., Jackson, M. J., & Ahmed, W., (2007). Titanium and titanium alloy applications in medicine. *International Journal of Nano- and Biomaterials, 1*(1), 3–34.
6. Bet-Moushoul, E., Mansourpanah, Y., Farhadi, K., & Tabatabaei, M. J. C. E. J., (2016). TiO_2 nanocomposite based polymeric membranes: A review on performance improvement for various applications in chemical engineering processes. *Chemical Engineering Journal, 283*, 29–46.
7. Bewer, G., Debrodt, H., & Herbst, H., (1982). Titanium for electrochemical processes. *JOM, 34*(1), 37–41.
8. Bhardwaj, S. K., Bhardwaj, N., Kaur, R., Mehta, J., Sharma, A. L., Kim, K. H., & Deep, A., (2018). An overview of different strategies to introduce conductivity in metal–organic frameworks and miscellaneous applications thereof. *Journal of Materials Chemistry A, 6*(31), 14992–15009.
9. Boyer, R. R., (1996). An overview on the use of titanium in the aerospace industry. *Materials Science and Engineering: A, 213*(1–2), 103–114.
10. Boyer, R. R., (2010). Attributes, characteristics, and applications of titanium and its alloys. *Jom, 62,* 21–24.
11. Bußmann, A. B., Durasiewicz, C. P., Kibler, S. H. A., & Wald, C. K., (2021). Piezoelectric titanium based microfluidic pump and valves for implantable medical applications. *Sensors and Actuators A: Physical, 323*, 112649.
12. Candeago, R., Kim, K., Vapnik, H., Cotty, S., Aubin, M., Berensmeier, S., & Su, X., (2020). Semiconducting polymer interfaces for electrochemically assisted mercury remediation. *ACS Applied Materials and Interfaces,*

12(44), 49713–49722.

13. Chen, X., & Deng, H., (2012). Removal of humic acids from water by hybrid titanium-based electrocoagulation with ultrafiltration membrane processes. *Desalination, 300*, 51–57.
14. Chin, Y. L., Chou, J. C., Lei, Z. C., Sun, T. P., Chung, W. Y., & Hsiung, S. K., (2001). Titanium nitride membrane application to extended gate field effect transistor pH sensor using VLSI technology. *Japanese Journal of Applied Physics, 40*(11R), 6311.
15. Clark, E. B., & Roebuck, B., (1992). Extending the application areas for titanium carbonitride cermets. *International Journal of Refractory Metals and Hard Materials, 11*(1), 23–33.
16. Clerici, M. G., (2015). The activity of titanium silicalite-1 (TS-1): Some considerations on its origin. *Kinetics and Catalysis, 56*(4), 450–455.
17. Darwish, W., Seikel, E., Käsmarker, R., Harms, K., & Sundermeyer, J., (2011). Synthesis and X-ray crystal structures of imido and urea to derivatives of titanium (IV) phthalocyanine and their application in the catalytic formation of carbodiimides by metathesis from isocyanates. *Dalton Transactions, 40*(8), 1787–1794.
18. David, O., Gendel, Y., & Wessling, M., (2014). Tubular macro-porous titanium membranes. *Journal of Membrane Science, 461*, 139–145.
19. Dong, H., (2010). Tribological properties of titanium-based alloys. In: *Surface Engineering of Light Alloys* (pp. 58–80). Woodhead Publishing.
20. Economou, A., & Fielden, P. R., (2003). Mercury film electrodes: Developments, trends, and potentialities for electroanalysis. *Analyst, 128*(3), 205–213.
21. Ehsani, M., & Aroujalian, A., (2020). Fabrication of electrospun polyethersulfone/titanium dioxide (PES/TiO_2) composite nanofibers membrane and its application for photocatalytic degradation of phenol in aqueous solution. *Polymers for Advanced Technologies, 31*(4), 772–785.
22. Eisenbarth, E., Velten, D., Müller, M., Thull, R., & Breme, J., (2004). Biocompatibility of β-stabilizing elements of titanium alloys. *Biomaterials, 25*(26), 5705–5713.
23. Elrod, C. W., (1980). The combustion of titanium in gas turbine engines. In: *Tri-Service Conference on Corrosion* (Vol. 2, pp. 55–92).
24. Fan, Y., Zhang, C., Liu, X., Lin, Y., Gao, G., Ma, C., & Li, X., (2019). Structure and transport properties of titanium oxide (Ti_2O, $TiO_1 + \delta$, and Ti_3O_5) thin films. *Journal of Alloys and Compounds, 786*, 607–613.
25. Fray, D. J., (2008). Novel methods for the production of titanium.

International Materials Reviews, 53(6), 317–325.

26. Freese, H. L., Volas, M. G., & Wood, J. R., (2001). Metallurgy and technological properties of titanium and titanium alloys. *Titanium in Medicine*, 25.
27. Froes, F. H., (2015). *Applications of Titanium* (pp. 353–379). Titanium physical metallurgy processing and applications, Materials Park, ASM International.
28. Gao, X., Zhou, Y., Tan, Y., Cheng, Z., Yang, B., Ma, Y., & Jia, J., (2019). Exploring adsorption behavior and oxidation mechanism of mercury on monolayer Ti_2CO_2 (MXenes) from first principles. *Applied Surface Science, 464*, 53–60.
29. Gerdemann, S. J., (2001). TITANIUM: Process technologies. *Advanced Materials and Processes, 159*(7), 41–41.
30. Ghorishi, S. B., & Sedman, C. B., (1998). Low concentration mercury sorption mechanisms and control by calcium-based sorbents: application in coal-fired processes. *Journal of the Air and Waste Management Association, 48*(12), 1191–1198.
31. Guizard, C., (1996). Sol-gel chemistry and its application to porous membrane processing. In: *Membrane Science and Technology* (Vol. 4, pp. 227–258). Elsevier.
32. Hantanasirisakul, K., Alhabeb, M., Lipatov, A., Maleski, K., Anasori, B., Salles, P., & Gogotsi, Y., (2019). Effects of synthesis and processing on optoelectronic properties of titanium carbonitride MXene. *Chemistry of Materials, 31*(8), 2941–2951.
33. Hayat, M. D., Singh, H., He, Z., & Cao, P., (2019). Titanium metal matrix composites: An overview. *Composites Part A: Applied Science and Manufacturing, 121*, 418–438.
34. Henriques, V. A. R., Campos, P. P. D., Cairo, C. A. A., & Bressiani, J. C., (2005). Production of titanium alloys for advanced aerospace systems by powder metallurgy. *Materials Research, 8*, 443–446.
35. Hu, B., Geng, F., Zhao, C., Doumert, B., Trébosc, J., Lafon, O., & Hu, B., (2020). Deciphering the origin of high electrochemical performance in a novel Ti-substituted P_2/O_3 biphasic cathode for sodium-ion batteries. *ACS Applied Materials and Interfaces, 12*(37), 41485–41494.
36. Hu, D., Dolganov, A., Ma, M., Bhattacharya, B., Bishop, M. T., & Chen, G. Z., (2018). Development of the Fray-Farthing-Chen Cambridge process: Towards the sustainable production of titanium and its alloys. *Jom, 70*, 129–137.

37. Hummer, D. R., Kubicki, J. D., Kent, P. R., Post, J. E., & Heaney, P. J., (2009). Origin of nanoscale phase stability reversals in titanium oxide polymorphs. *The Journal of Physical Chemistry C, 113*(11), 4240–4245.
38. Ikeda, B. M., & Shoesmith, D. W., (1997). *Industrial Experience with Titanium (No. AECL-11750)*. Atomic Energy of Canada Limited.
39. Jiang, Y., Deng, C., He, Y., Zhao, Y., Xu, N., Zou, J., & Liu, C. T., (2009). Reactive synthesis of microporous titanium-aluminide membranes. *Materials Letters, 63*(1), 22–24.
40. Kang, X., Liu, S., Dai, Z., He, Y., Song, X., & Tan, Z., (2019). Titanium dioxide: From engineering to applications. *Catalysts, 9*(2), 191.
41. Khor, K. A., Gu, Y. W., Quek, C. H., & Cheang, P., (2003). Plasma spraying of functionally graded hydroxyapatite/Ti-6Al-4V coatings. *Surface and Coatings Technology, 168*(2–3), 195–201.
42. Khorasani, A. M., Goldberg, M., Doeven, E. H., & Littlefair, G., (2015). Titanium in biomedical applications – properties and fabrication: A review. *Journal of Biomaterials and Tissue Engineering*, 5(8), 593–619.
43. Kubaschewski, O., De Keyzer, J., Schmid-Fetzer, R., Shcherban, O., Tomashik, V., Jialin, Y., & Tretyachenko, L., (2007). *Silver-Copper-Titanium*.
44. Lakshmanan, V. I., Sridhar, R., Sheikhzeinoddin, T., Halim, M. A., & Roy, R., (2012). Extraction of titanium and vanadium by chloride leach processes. In: *TT Chen Honorary Symposium on Hydrometallurgy, Electrometallurgy, and Materials Characterization, TMS (The Minerals, Metals, and Materials Society)* (pp. 295–301).
45. Li, B. Y., Rong, L. J., Li, Y. Y., & Gjunter, V. E., (2000). A recent development in producing porous Ni–Ti shape memory alloys. *Intermetallics, 8*(8), 881–884.
46. Livraghi, S., Paganini, M. C., Giamello, E., Selloni, A., Di Valentin, C., & Pacchioni, G., (2006). Origin of photoactivity of nitrogen-doped titanium dioxide under visible light. *Journal of the American Chemical Society, 128*(49), 15666–15671.
47. Loria, E. A., (2000). Gamma titanium aluminides as prospective structural materials. Intermetallics, 8(9–11), 1339–1345.
48. Ma, M., Wang, D., Wang, W., Hu, X., Jin, X., & Chen, G. Z., (2006). Extraction of titanium from different titania precursors by the FFC Cambridge process. *Journal of Alloys and Compounds, 420*(1–2), 37–45.
49. Maykuth, D. J., & Wood, R. A., (1969). *Nonaerospace Uses of Titanium* (Vol. 242). Battelle Memorial Institute, Defense Metals Information

Center.

50. Moeller, T., (1950). Titanium: Its occurrence, chemistry, and technology. Jelks Barksdale. New York: Ronald Press, 1949. 591 pp. $10.00. *Science, 111*(2886), 418–418.
51. Noel, J. D., Biswas, P., & Giammar, D. E., (2007). Evaluation of a sequential extraction process used for determining mercury binding mechanisms to coal combustion byproducts. *Journal of the Air and Waste Management Association, 57*(7), 856–867.
52. Notari, B., (1988). Synthesis and catalytic properties of titanium containing zeolites. In: *Studies in Surface Science and Catalysis* (Vol. 37, pp. 413–425). Elsevier.
53. Ohno, T., Haga, D., Fujihara, K., Kaizaki, K., & Matsumura, M., (1997). Unique effects of iron(III) ions on photocatalytic and photoelectrochemical properties of titanium dioxide. *The Journal of Physical Chemistry B, 101*(33), 6415–6419.
54. Ouchi, C., Fukai, H., & Hasegawa, K., (1999). Microstructural characteristics and unique properties obtained by solution treating or aging in β-rich α+ β titanium alloy. *Materials Science and Engineering: A, 263*(2), 132–136.
55. Pan, Y., Wang, T., Sun, H., & Wang, W., (2012). Preparation and application of titanium dioxide dynamic membranes in microfiltration of oil-in-water emulsions. *Separation and Purification Technology, 89*, 78–83.
56. Panigrahi, M., Paramguru, R. K., Gupta, R. C., Shibata, E., & Nakamura, T., (2010). An overview of production of titanium and an attempt to titanium production with ferro-titanium. *High Temperature Materials and Processes, 29*(5–6), 495–514.
57. Parrino, F., Pomilla, F. R., Camera-Roda, G., Loddo, V., & Palmisano, L., (2021). Properties of titanium dioxide. In: *Titanium Dioxide (TiO) and its Applications* (pp. 13–66). Elsevier.
58. Pelegrini, R. T., Freire, R. S., Duran, N., & Bertazzoli, R., (2001). Photo-assisted electrochemical degradation of organic pollutants on a DSA type oxide electrode: process test for a phenol synthetic solution and its application for the E1 bleach Kraft mill effluent. *Environmental Science and Technology, 35*(13), 2849–2853.
59. Perego, C., Carati, A., Ingallina, P., Mantegazza, M. A., & Bellussi, G., (2001). Production of titanium containing molecular sieves and their application in catalysis. *Applied Catalysis A: General, 221*(1–2), 63–72.
60. Poletti, C., Balog, M., Schubert, T., Liedtke, V., & Edtmaier, C., (2008). Production of titanium matrix composites reinforced with SiC particles.

Composites Science and Technology, 68(9), 2171–2177.

61. Pourbaix, M., (1974). Applications of electrochemistry in corrosion science and in practice. *Corrosion Science, 14*(1), 25–82.
62. Razmjou, A., Arifin, E., Dong, G., Mansouri, J., & Chen, V., (2012). Superhydrophobic modification of TiO_2 nanocomposite PVDF membranes for applications in membrane distillation. *Journal of Membrane Science, 415*, 850–863.
63. Rust, T. M., Seth, B. B., & Thompson, T. L., (1990). Application of titanium to low pressure transition stage steam turbine blading at Tu Electric. In: *Titanium Steam Turbine Blading* (pp. 169–214). Pergamon.
64. Santos, G., (2017). The importance of metallic materials as biomaterials. *Adv. Tissue Eng. Regen. Med. Open Access, 3*(1), 300–302.
65. Setoyama, D., Matsunaga, J., Muta, H., Uno, M., & Yamanaka, S., (2004). Mechanical properties of titanium hydride. *Journal of alloys and Compounds, 381*(1–2), 215–220.
66. Shi, W., Shen, J., Lu, S., Chen, B., Atif, M., & Li, Y., (2023). Origin of the excellent ductility in super high oxygen doped titanium. *Materials Science and Engineering: A, 862*, 144429.
67. Singh, A., Ramachandran, S. K., Gumpu, M. B., Zsuzsanna, L., Veréb, G., Kertész, S., & Gangasalam, A., (2021). Titanium dioxide doped hydroxyapatite incorporated photocatalytic membranes for the degradation of chloramphenicol antibiotic in water. *Journal of Chemical Technology and Biotechnology, 96*(4), 1057–1066.
68. Sioshansi, P., (1987). Medical applications of ion beam processes. *Nuclear Instruments and Methods in Physics Research Section B: Beam Interactions with Materials and Atoms, 19*, 204–208.
69. Tak, S., & Kumar, A., (2017). Chlorination disinfection by-products and comparative cost analysis of chlorination and UV disinfection in sewage treatment plants: Indian scenario. *Environmental Science and Pollution Research, 24*, 26269–26278.
70. Tan, C. W., & Miao, J., (2009). Optimization of sputtered Cr/Au thin film for diaphragm-based MEMS applications. *Thin Solid Films, 517*(17), 4921–4925.
71. Thompson, S. A., (2000). An overview of nickel–titanium alloys used in dentistry. *International Endodontic Journal, 33*(4), 297–310.
72. Tsai, C. Y., Liu, C. W., Hsi, H. C., Lin, K. S., Lin, Y. W., & Lai, L. C., (2019). Synthesis of Ag-modified TiO_2 nanotube and its application in photocatalytic degradation of dyes and elemental mercury. *Journal of*

Chemical Technology and Biotechnology, 94(10), 3251–3262.

73. Turkevych, I., Pihosh, Y., Goto, M., Kasahara, A., Tosa, M., Kato, S., & Koguchi, N., (2008). Photocatalytic properties of titanium dioxide sputtered on a nanostructured substrate. *Thin Solid Films, 516*(9), 2387–2391.
74. Ullman, A., (2008). Cost saving in chlorine plants by benefitting from the unique properties of titanium. *Modern Chlor-Alkali Technology, 8*, 282.
75. Vieira, T. M., Sagás, J. C., Pessoa, R. S., & Fontana, L. C., (2020). Multiphase titanium suboxides films grown on glass substrate: Surface energy, wettability, and morphology analysis. *Surfaces and Interfaces, 20*, 100610.
76. Welsch, G., Boyer, R., & Collings, E. W., (1993). *Materials Properties Handbook: Titanium Alloys*. ASM international.
77. Williams, J. C., & Boyer, R. R., (2020). Opportunities and issues in the application of titanium alloys for aerospace components. *Metals, 10*(6), 705.
78. Xie, J., Yan, N., Yang, S., Qu, Z., Chen, W., Zhang, W., & Jia, J., (2012). Synthesis and characterization of nano-sized Mn–TiO_2 catalysts and their application to removal of gaseous elemental mercury. *Research on Chemical Intermediates, 38*, 2511–2522.
79. Yang, S., Guo, Y., Yan, N., Wu, D., He, H., Qu, Z., & Jia, J., (2011). Nanosized cation-deficient Fe–Ti spinel: a novel magnetic sorbent for elemental mercury capture from flue gas. *ACS Applied Materials and Interfaces, 3*(2), 209–217.

INDEX

D

E

F

G

H

I

L

M

N

Printed in the United States
by Baker & Taylor Publisher Services